CHEMISTRY RESEARCH AND APPLICATIONS

A COMPREHENSIVE GUIDE TO NATURAL PRODUCTS

CHEMISTRY RESEARCH AND APPLICATIONS

Additional books and e-books in this series can be found on Nova's website under the Series tab.

CHEMISTRY RESEARCH AND APPLICATIONS

A COMPREHENSIVE GUIDE TO NATURAL PRODUCTS

SILJE A. DAHL
AND
ADAM M. FRANDSEN
EDITORS

Copyright © 2020 by Nova Science Publishers, Inc.

All rights reserved. No part of this book may be reproduced, stored in a retrieval system or transmitted in any form or by any means: electronic, electrostatic, magnetic, tape, mechanical photocopying, recording or otherwise without the written permission of the Publisher.

We have partnered with Copyright Clearance Center to make it easy for you to obtain permissions to reuse content from this publication. Simply navigate to this publication's page on Nova's website and locate the "Get Permission" button below the title description. This button is linked directly to the title's permission page on copyright.com. Alternatively, you can visit copyright.com and search by title, ISBN, or ISSN.

For further questions about using the service on copyright.com, please contact:
Copyright Clearance Center
Phone: +1-(978) 750-8400 Fax: +1-(978) 750-4470 E-mail: info@copyright.com.

NOTICE TO THE READER

The Publisher has taken reasonable care in the preparation of this book, but makes no expressed or implied warranty of any kind and assumes no responsibility for any errors or omissions. No liability is assumed for incidental or consequential damages in connection with or arising out of information contained in this book. The Publisher shall not be liable for any special, consequential, or exemplary damages resulting, in whole or in part, from the readers' use of, or reliance upon, this material. Any parts of this book based on government reports are so indicated and copyright is claimed for those parts to the extent applicable to compilations of such works.

Independent verification should be sought for any data, advice or recommendations contained in this book. In addition, no responsibility is assumed by the Publisher for any injury and/or damage to persons or property arising from any methods, products, instructions, ideas or otherwise contained in this publication.

This publication is designed to provide accurate and authoritative information with regard to the subject matter covered herein. It is sold with the clear understanding that the Publisher is not engaged in rendering legal or any other professional services. If legal or any other expert assistance is required, the services of a competent person should be sought. FROM A DECLARATION OF PARTICIPANTS JOINTLY ADOPTED BY A COMMITTEE OF THE AMERICAN BAR ASSOCIATION AND A COMMITTEE OF PUBLISHERS.

Additional color graphics may be available in the e-book version of this book.

Library of Congress Cataloging-in-Publication Data

ISBN: 978-1-53618-418-1
Names: Dahl, Silje A., editor.
Title: A comprehensive guide to natural products / Silje A. Dahl and Adam M. Frandsen, (editors).
Description: New York : Nova Science Publishers, [2020] | Series: Chemistry research and applications | Includes bibliographical references and index. |
Identifiers: LCCN 2020032611 (print) | LCCN 2020032612 (ebook) | ISBN 9781536184181 (paperback) | ISBN 9781536184327 (adobe pdf)
Subjects: LCSH: Botanical chemistry. | Phytochemicals. | Natural products--Biotechnology.
Classification: LCC QK865 .C66 2020 (print) | LCC QK865 (ebook) | DDC 572/.2--dc23
LC record available at https://lccn.loc.gov/2020032611
LC ebook record available at https://lccn.loc.gov/2020032612

Published by Nova Science Publishers, Inc. † New York

CONTENTS

Preface		vii
Chapter 1	Trends of Liquid Chromatography, Mass Spectrometry and Chemometrics in the Analysis and Characterization of Plant Natural Products *Nerea Núñez, Laura Cayero, Aina Cuixart-Gimó and Oscar Núñez*	1
Chapter 2	Pharmacological Interest of Cetrarioid Lichens in the Prevention of Oxidative Stress-Related Diseases *Isabel Ureña-Vacas, Elena González-Burgos and M. Pilar Gómez-Serranillos*	91
Chapter 3	Phytochemicals: An Approach towards Antisickling Activity *Jaya Tiwari, Vijaylakshmi Jain and Pankaj Kishor Mishra*	129
Chapter 4	Phytochemical Potency and Biocompatible Comparison of Green and Organic Solvent *Vijaylakshmi Jain, Jaya Tiwari and Pankaj Kishor Mishra*	149

Chapter 5	Phytochemical Activity Against Drug-Resistant Microbes: Current Status and Future Prospects *Md. Didaruzzaman Sohel and Andrew W. Taylor-Robinson*	**169**
Index		**187**

Preface

In this compilation, the role of liquid chromatography, mass spectrometry and chemometrics for the analysis and characterization of plant natural products is addressed.

The authors provide a comprehensive review of the pharmacological activity of cetrarioid lichens and their major secondary metabolites as antioxidants to prevent and treat oxidative stress-related diseases.

Following this, the way in which the detection of various secondary metabolites and bioactive compounds in some plants can reduce sickle cells in vitro is studied.

In addition, the efficiency of green and conventional solvent systems concerning the three classes of phytochemicals (phenols, alkaloids, and flavonoids) is described.

Lastly, a brief history of antibiotics and the spread of resistance is provided, and future strategies to combat drug-resistant microbes are discussed.

Chapter 1 - A natural product is any organic phytochemical that is synthesized by a living organism. Today, the growing interest in the characterization and analysis of natural products of plant, bacterial or fungi origin is unquestionable due to their potential applications as new drugs, functional foods, nutraceuticals or bioactive compounds. Natural products have been studied for the prevention of many diseases, including cancer.

More than 800 new approved drugs in the last years were of natural origin. Analytical methodologies for the isolation, identification, and characterization of natural products are on demand.

Liquid chromatography (LC) separation methodologies are nowadays among the most employed strategies to address the analysis and characterization of natural products. The combination of these LC techniques with low-resolution mass spectrometry (LC-MS) and tandem mass spectrometry (LC-MS/MS), either using ion trap (IT), linear ion trap (LIT), and triple quadrupole (QqQ) instruments, are the most frequently proposed methodologies for the determination of plant-based natural products. Of great importance in this field is the combination of liquid chromatography techniques with high-resolution mass spectrometry (LC-HRMS), mainly using time-of-flight and Orbitrap mass analyzers. The high-resolution power and accurate mass determination achieved with these techniques are crucial tools for the characterization and identification of new natural products in very complex matrices such as plants with a huge number of phytochemicals with grate differences in properties, structures, and concentration levels. Obviously, due to the high number of chemical data obtained by these techniques, chemometrics becomes an indispensable tool for the correct data processing. In this chapter, the role of liquid chromatography, mass spectrometry and chemometrics for the analysis and characterization of plant natural products will be addressed. Coverage of all kinds of applications in such an important field is beyond the scope of the present contribution, so the chapter will focus on the most relevant applications published in the last years.

Chapter 2 - Lichens are symbiotic organisms composed by a mycobiont (fungus) and a photobiont (unicellular algae or cyanobacteria). Recent studies have revealed multispecific symbiosis with host-specific bacterial microbiomes. It is estimated that there are more than 28,000 lichen species worldwide among which Lecanorales order being the most abundant and including as the largest family Parmealiaceae, highlighting within it Cetrarioid clade.

Oxidative stress occurs when there is a disturbance between reactive oxygen species (ROS) and the antioxidant system (enzymatic and non-enzymatic), which may contribute to pathological chronic conditions including cancer, cardiovascular diseases and neurodegenerative disorders. Exogenous antioxidants administration is considered the most promising strategy to cope oxidative stress based on its capacity to inhibit ROS action, chelate metal ions and increase enzyme activity and expression. At present, the interest in searching natural compounds as new antioxidants is increasing. Experimental studies have demonstrated that cetrarioid lichens and its major secondary metabolites are potential antioxidants with interest to prevent oxidative stress related diseases.

This chapter aims to provide a comprehensive overview of the pharmacological activity of cetrarioid lichens and its major secondary metabolites as antioxidants to prevent and treat oxidative stress-related diseases.

Chapter 3 - Sickle cell disease (SCD) is the most prevalent inherited blood disorder affecting most parts of the world without any discrimination of ethics or racial standards. The patients undergo shortness of breath, heart palpitation, abdominal and muscle pain. Several managing SCD therapies have been proposed with treatment but all these treatments are ineffective or very expensive for the less fortunate population. Plant species used as folk medicines also display *in vitro* antisickling activity. Caricaceae (*Carica papaya*), Fabaceae (*Cajanus cajan* and *Crotalaria retusa*), Apocynaceae (*Raulwolfia vomitoria, Jatropha curcas, Euphorbia hirta* and *Picralima nitida*) and Euphorbiaceae (*Wrightia tinctoria* and *Alchornea cordifolia*) are prominently studied families of plants and their parts with their respective solvents for antisickling activity. Detection of various secondary metabolites like alkaloid, flavonoids, tannin, anthraquinone and many more bioactive compounds in these plants can be eminent cause for reducing sickle cells *in vitro* and can be further used for the development of therapeutic agent for cure of disease.

Chapter 4 - Medicinal plants are the repository of therapeutical drugs in terms of secondary metabolite or phytochemicals. The extraction of this secondary metabolite is one of the major steps of the whole process. Solvents despite being part of the whole process are not part of the composition and formulation process. Protracted exposure to solvents has adverse effects on all living organisms, causing detrimental effects on the body. Reduced use or replacement of such solvent with less toxicity and the harmful effect is the need of today. Green technology or green solvent systems has brought a revolutionary change in the toxin world. A less toxic, easy accessibility, higher possibility of reuse and a great efficient approach is how green solvent is characterized. Solvent efficiency is highly impacted by the type of solvent and so the phytochemical constituent. This chapter describes the efficiency and comparative analysis of green and conventional solvent systems concerning the three classes of phytochemicals i.e., phenols, alkaloids, and flavonoids. It also accounts for the antimicrobial activity against gram-negative, gram-positive and fungal strains of both the solvent system and its efficiency. The implications of the findings for phytochemicals show that extraction efficiency of phytochemicals was found high in the green solvent than organic ones and antimicrobial activity was also observed high in organic than the conventional one. The study discusses the solvent, solvent system, phytochemicals, antimicrobial infections and future perspective to the whole outcome.

Chapter 5 - The advance of antimicrobial resistance to existing frontline therapeutics is widely recognized as a global health threat. In order to address the increasing challenge that this presents to patient treatment by the medical profession pharmaceutical and biotechnology sectors continue to seek novel therapeutic agents to which pathogens, notably bacteria, are sensitive.

The discovery of antibiotics in the last century led to a rapid and profound reduction in morbidity and mortality associated with commonly occurring bacterial diseases. However, the ongoing heavy reliance and indiscriminate use has resulted, due to genetic mutation under selective pressure, in the emergence of antibiotic-resistant bacteria. In searching for new commercial sources of antimicrobials crude extracts of medicinal plants have attracted attention. The wide range of metabolites – for example, alkaloids, tannins and polyphenols – carry therapeutic potential as either novel antimicrobials or modifiers of existing resistance. Plant extracts containing such phytochemicals are able to bind to protein domains, thereby modifying or inhibiting protein-protein interactions. This enables these herbal derivatives to act as effective modulators of cellular metabolic pathways involved in the immune response, mitosis, apoptosis and signal transduction. Hence, the mechanism(s) of action may not necessarily be directly microbicidal but instead affect key events within the host cell that reduce the ability of bacteria, fungi and viruses to thrive in an intracellular environment.

In this chapter, a brief history of antibiotics and the spread of resistance is provided. The authors describe phytochemicals that are currently known and outline their antimicrobial activities. Future strategies to combat drug-resistant microbes are discussed.

In: A Comprehensive Guide ...
Editors: Silje A. Dahl et al.

ISBN: 978-1-53618-418-1
© 2020 Nova Science Publishers, Inc.

Chapter 1

TRENDS OF LIQUID CHROMATOGRAPHY, MASS SPECTROMETRY AND CHEMOMETRICS IN THE ANALYSIS AND CHARACTERIZATION OF PLANT NATURAL PRODUCTS

Nerea Núñez[1], Laura Cayero[1], Aina Cuixart-Gimó[1] and Oscar Núñez[1,2,3,]*

[1]Department of Chemical Engineering and Analytical Chemistry, University of Barcelona, Barcelona, Spain
[2]Research Institute in Food Nutrition and Food Safety, University of Barcelona, Barcelona, Spain
[3]Serra Húnter Fellow, Generalitat de Catalunya, Spain

ABSTRACT

A natural product is any organic phytochemical that is synthesized by a living organism. Today, the growing interest in the characterization and

---

* Corresponding Author's Email: oscar.nunez@ub.edu.

analysis of natural products of plant, bacterial or fungi origin is unquestionable due to their potential applications as new drugs, functional foods, nutraceuticals or bioactive compounds. Natural products have been studied for the prevention of many diseases, including cancer. More than 800 new approved drugs in the last years were of natural origin. Analytical methodologies for the isolation, identification, and characterization of natural products are on demand.

Liquid chromatography (LC) separation methodologies are nowadays among the most employed strategies to address the analysis and characterization of natural products. The combination of these LC techniques with low-resolution mass spectrometry (LC-MS) and tandem mass spectrometry (LC-MS/MS), either using ion trap (IT), linear ion trap (LIT), and triple quadrupole (QqQ) instruments, are the most frequently proposed methodologies for the determination of plant-based natural products. Of great importance in this field is the combination of liquid chromatography techniques with high-resolution mass spectrometry (LC-HRMS), mainly using time-of-flight and Orbitrap mass analyzers. The high-resolution power and accurate mass determination achieved with these techniques are crucial tools for the characterization and identification of new natural products in very complex matrices such as plants with a huge number of phytochemicals with grate differences in properties, structures, and concentration levels. Obviously, due to the high number of chemical data obtained by these techniques, chemometrics becomes an indispensable tool for the correct data processing. In this chapter, the role of liquid chromatography, mass spectrometry and chemometrics for the analysis and characterization of plant natural products will be addressed. Coverage of all kinds of applications in such an important field is beyond the scope of the present contribution, so the chapter will focus on the most relevant applications published in the last years.

Keywords: natural products, liquid chromatography, low-resolution mass spectrometry, high-resolution mass spectrometry, chemometrics

INTRODUCTION

Natural products are chemical compounds or substances produced by any living organism, including plants, fungi, bacteria and marine organisms [1]. Although the term natural product is nowadays also extended, for commercial purposes, to cosmetics, dietary supplements, and foods produced from natural sources without the addition of any artificial ingredient, from the point of view of the field of organic chemistry, natural products are restricted to purified phytochemicals isolated from natural sources. These compounds are produced by the pathway of primary or secondary metabolism of living organisms [2]. Primary metabolites such as amino acids, lipids, and carbohydrates, are necessary for physiology purposes and are directly involved in the normal growth, development and reproduction of living organisms [3, 4]. In contrast, secondary metabolites are nonessential to sustain the life of a given organism, as they are not directly involved in their growth, development and reproduction, but they have an ecological function becoming necessary for the living organism survival in a given environment [3, 5]. These secondary metabolites are specialized chemicals synthetized by living organisms such as plants and microorganisms, and they are involved in processes like defense against other organisms or environmental situations, or used as attractants for reproduction purposes [6]. Due to the high diversity and availability, plants are among the most important source of these secondary metabolites. Among them, the most bioactive ones are alkaloids, tannins, carotenoids, flavonoids, phenolic compounds, peptides and sterols [7, 8]. Regarding the plant kingdom, secondary metabolites also differ from primary metabolites in their restricted distribution. While primary metabolites are found throughout all the plant kingdom, secondary metabolites tend to be found in only one plant species or related group of species, and in specific tissues like the leaves, stem, root or the bark of the plant depending on the type of secondary metabolite that is been synthesized [7]. Since the nineteenth and early twentieth centuries, society is interested in these substances due to their importance as medicinal drugs, poisons, flavors and industrial materials [9]. As a result, natural product research is today one of the most important fields concerned in identifying pharmacologically active compounds from natural sources, especially from plant-based tissues.

The importance and the need in research to discover new plant-based natural products is indisputable, due to the possible benefits in the prevention of many diseases, including cancer [10]. This is revealed by the fact that more than 800 new approved drugs in the last years were of natural origin [11], and 14 natural product-based drugs were approved for marketing worldwide between 2005 and 2014 [12]. Besides, taking into account the great diversity of plant species, and consequently plant-derived products potentially available, and the potential applications of the new-discovered bioactive compounds, the development of efficient analytical methodologies for the fast chemical and biological screening of plant and plant-based product extracts is necessary. And this is clearly observed with the increasing number of publications in this area.

However, the analysis of plant and plant-based product extracts (fruits, vegetables, etc.) for the isolation and identification of new bioactive natural products is not an easy task. This is due to the fact that plant extracts usually contain hundreds of chemical constituents, most of them small molecules within the group of secondary metabolites. For example, only the polyphenolic compounds, aromatic secondary metabolites ubiquitously spread through the plant kingdom, constitute already a group of chemicals comprising more than 8,000 substances [13]. Also, the range of molecular masses that can be found within this huge group of plant chemical constitutes goes from low small molecules (~50 – 100 Da), such as the case of some phenolic acids, medium size molecules (~500 – 6,000 Da) such as peptides, to big molecules (>30,000 Da) of highly polymerized compounds. Besides, these compounds may have remarkably different physicochemical properties (solubility, polarities, bioavailability, etc.), as well as different structures. And last but not least, the amounts at which these plant chemical constituents are found are highly variable, from relatively high concentrations (hundreds of mg/L) to trace levels (µg/L, ng/L, or even lower). And in addition, in many cases, the compounds of greatest interest to the pharmaceutical industry for their potential disease-preventing properties may be those found at lower concentrations. Therefore, the separation and characterization of plant chemical constituents as natural products is a challenging task in the analytical and separation sciences.

Liquid chromatography (LC)-based separation methodologies are nowadays among the most employed strategies for the analysis and characterization of natural products, including high-performance liquid chromatography (HPLC), ultra-high-performance liquid chromatography (UHPLC), and nano liquid chromatography (nano-LC). Apart from the use of ultraviolet (UV) or fluorescence (FL) detection, the combination of these LC techniques with low-resolution mass spectrometry (LRMS), in both MS (LC-MS) and tandem mass spectrometry (LC-MS/MS) modes, is also widely employed for the determination of bioactive natural products. In these applications, triple quadrupole (QqQ), ion trap (IT) and linear ion trap (LIT) instruments are typically employed as the analyzers [14–17]. Frequently, LRMS methodologies are employed for the determination of previously identified and isolated plant-based natural products, either for quantitation purposes or for the study of fragmentation pathways in the case of ion-trap-based analyzers.

Of very great importance in this field is the combination of liquid chromatography techniques with high-resolution mass spectrometry (LC-HRMS), methodologies typically employed for the characterization of natural products with the aim of identifying new bioactive compounds. Time-of-flight (TOF) and Orbitrap instruments, as well as hybrid combinations such as quadrupole-time-of-flight (Q-TOF), quadrupole-Orbitrap (Q-Orbitrap), or linear trap quadrupole-Orbitrap (LTQ-Orbitrap), are among the most frequently HRMS analyzers employed for that purpose [4, 8, 16]. These instruments provide a wide variety of strategies for the characterization and identification of new natural products thanks to their potentially unlimited m/z rage and high-resolution power while keeping high sensitivity and mass accuracy [18, 19]. These are important instrument characteristics to address the identification of specific phytochemicals in so complex mixtures such as plant-based products with a huge number of compounds differing in physicochemical properties, structures and concentration levels, as previously commented. Besides, these instruments have contributed greatly in the development of the "omics" techniques (e.g., proteomics, metabolomics, foodomics). In fact, metabolomics refers to the analysis of small-molecular-weight molecules (typically <1,500 Da), where many potential plant-based bioactive secondary metabolites can be found. Recently, two dimensional liquid chromatography (2DLC) in combination with mass spectrometry is also being employed for the characterization and chemical analysis of plant natural products [20, 21].

Obviously, due to the high number of chemical data obtained by the commented techniques, chemometrics [22, 23], the science of extracting information from chemical systems by data-driven means, becomes an indispensable tool for the correct data processing. Also, combination of LC-MS or LC-HRMS with other separation techniques such as gas chromatography (GC) and capillary electrophoresis (CE), or with other characterization techniques such as nuclear magnetic resonance (NMR), as well as the use of *in Silico* databases, together with chemometrics [4], is also very frequent in order to be able to identify small amounts of these secondary metabolites often produced in small quantities, in a background of known metabolites present in higher quantities.

This chapter aims to address the role of liquid chromatography, mass spectrometry and chemometrics for the analysis and characterization of plant natural products. Obviously, coverage of all kinds of published applications in such as important field is beyond the scope of the present contribution. The chapter will focus on the most relevant applications published since 2015.

LIQUID CHROMATOGRAPHY TECHNIQUES

Liquid chromatography (LC) is the most employed technique for the qualitative and quantitative analysis of natural products. LC presents many advantages over other separation techniques. For example, it allows the determination of a huge variety of chemical compounds (with different physicochemical properties) and working with a lot of different matrices by choosing the stationary phase composition (column) and the mobile phase components, according to both the analytes and the sample matrix. Besides, LC can work at different flow rates and temperatures, and can provide high resolution and performance. In addition, LC can be combined with different detection systems being able to provide high sensitivity depending on the application.

Clearly, liquid chromatography in combination with spectroscopy detection systems, i.e., ultraviolet (UV), fluorescence (FL), and nuclear magnetic resonance (NMR) detection, as well as with evaporative light scattering detection (ELSD), and mass spectrometry (MS), among other detection systems, makes liquid chromatography a powerful tool to identify natural product compounds and to study their biological activity and their structures [14].

Table 1 summarizes some selected LC methodologies with different detection systems for the analysis and characterization of plant natural products. As can be seen, reversed-phase liquid chromatography (RPLC) using C18 columns and gradient elution with an acidified aqueous solution and methanol or acetonitrile as mobile phase components is usually proposed [24–43]. For example, Hoj et al. [25] described the development of a precise and robust method for the quantification of free formaldehydes in commercial carrageenan and processed *Eucheuma* seaweed samples, using HPLC-UV, to guarantee their secure consumption. The separation was performed with a C18 column in 14 minutes, using gradient elution with water and methanol. In another work, Aysun et al. [28] described the use of C18 columns to guarantee the completely separation of phenolic compounds in the medicinal plant *Stachys pumila* extract in 70 minutes to achieve their quantification and study their antioxidant activity in front of hepatotoxicity induced by CCl_4^-. Other example of the use of C18 columns is described in the study by Al-Rimawi et al. [29] where they evaluated the content of flavonoids and phenolic compounds in three medicinal plants from Jordania to study their antibacterial activity. Both polar and non-polar compounds were eluted in combination with a separation achieved in less than 60 minutes. Some C18 column variations such as XDB-C18 reversed-phase columns were employed by Najafian et al. [35] for the quantification of the polyphenolic content in peppermint (*Mentha piperita*) and stevia (*Stevia rebaudiana*). The use of this column was very appropriate to work at pH around 3 with ionizable compounds. Gradient elution was also selected to achieve a satisfactory separation and a good selectivity.

Table 1. Selected LC methodologies with different detection systems for the analysis and characterization of plant natural products

Sample (compounds)	LC conditions	Chemometrics	Ref.
Potentilla anserina L. (phenolic compounds)	ProntoSIL-120-5-C18 AQ (60x1 mm I.D., 5 µm) Gradient elution (0.6 mL/min): (A) 0.2 m $LiClO_4$ in 0.006 M $HClO_4$, (B) acetonitrile UV-visible detection (λ 270 nm)	-	[24]
Carrageenan and processed Eucheuma (free formaldehyde)	Atlantis dC18 (150x4.6 mm I.D., 3 µm) Gradient elution (1 mL/min): water and methanol (40/60 v/v) UV-visible detection (λ 355 nm)	-	[25]
Punica granatum L. (phenolic compounds)	Luna C18 (250x4.6 mm I.D., 5 µm) Gradient elution (1 mL/min): (A) water with 5% formic acid, (B) methanol DAD (λ 280, 360 and 520 nm)	-	[36]
Ribes L. (Rutin and Chlorogenic Acid)	Waters UPLC BEH phenyl column (100x1.0 mm I.D., 1.7 µm) Gradient elution (0.3 mL/min): (A) acetonitrile and 0.1 M formic acid buffer, pH = 3.77 (15:85 v/v) in the presence of 1.0 mL triethylamine in 1000 mL mobile phase UV-visible detection (λ 290 to 360 nm)	PCR, PLS	[46]
Fritillariae Thunbergii Bulbus (Peimine and peiminine)	XBridge Shield RP 18 column (250x4.6 mm I.D., 3.5 µm) Gradient elution (1 mL/min): (A) water with 0.03% diethylamine, (B) acetonitrile UV-visible detection (λ 190 to 400 nm)	-	[37]
Rice bran of Aush Dhan (phenolic compounds)	Acclaim C18 (250x4.6 mm I.D., 5 µm) Gradient elution (1 mL/min): (A) acetonitrile, (B) acetic solution of pH 3.0, (B) methanol UV-visible detection (λ 200 to 700 nm)	-	[38]
Aloe trigonantha L.C. (Latex)	IB-Sil C18 (250x3.2 mm I.D., 5 µm) Gradient elution (1 mL/min): (A) water (B) methanol UV-visible detection (λ 254 nm) IR (400-4000 cm^{-1}) 1H (400 MHz) and 13C NMR (100 MHz) 2D-NMR	-	[39]
Descurainia sophia (L.) (fingerprinting)	Acclaim TM 120-C18 (250x4.6 mm I.D., 5 µm) Gradient elution (1 mL/min): (A) water with 1% phosphoric acid (B) acetonitrile	HCA, PCA, PLS-DA	[40]

Sample (compounds)	LC conditions	Chemometrics	Ref.
	UV-visible detection (λ 330 nm)		
Colubrina greggii (bioactive metabolites)	Gemini C18 (150x4.6 mm I.D., 5 µm) Gradient elution (1.5 mL/min): (A) water, (B) acetonitrile UV-visible detection (λ 210 nm)	PCA, O-PLS	[41]
Medicinal herbs (phenolic compounds)	Zorbax SB-Aq column (250x4.6 mm I.D., 5 µm) Gradient elution (1 mL/min): (A) water with 0.5% formic acid, (B) methanol UV-visible detection (λ 250, 280 and 320 nm)	-	[44]
Cassia fistula flower (phenolic compounds)	Water C18 column Gradient elution (1 mL/min): (A) methanol (B) water with 0.1% trifluoroacetic acid, (B) methanol UV-visible detection (λ 280 nm)	-	[42]
Stachys pumila (phenolic compounds)	Inertsil ODS 3 (250x4.6 mm I.D., 5 µm) Gradient elution (0.7 mL/min): (A) water with 2% acetic acid, (B) water and acetonitrile (50:50, *v/v*) with 0.5% acetic acid UV-visible detection (λ 190 to 800 nm)	-	[28]
Mentha piperita and *Stevia rebaudiana* (polyphenols)	Zorbax Eclipse XDB-C18 (150x10 mm I.D., 5 µm) Gradient elution: (A) water with 1% formic acid, (B) methanol UV-visible detection (λ 230 and 280 nm)	-	[35]
Chimonanthus salicifolius (fingerprint analysis)	Waters Acquity BEH Phenyl (50x2.1 mm I.D., 1.7 µm) Gradient elution (0.2 mL/min): (A) water with 0.1% phosphoric acid, (B) acetonitrile PAD (λ 190 to 400 and 210 nm)	HCA, PCA	[45]
Myristica fragans L. and *Pimenta dioica* (phenolic compounds)	Knauer RP-C18 (250x4.6 mm I.D., 5 µm) Gradient elution (1 mL/min): (A) water with 0.1% formic acid, (B) acetonitrile with 0.1% formic acid UV-visible detection (λ 254 and 280 nm)	-	[43]
Gentiana crassicaulis (iridoid glycosides)	Sepax Gp-C18 (150x4.6 mm I.D., 5 µm) Gradient elution (1 mL/min): (A) water with 0.1% formic acid, (B) methanol UV-visible detection (λ 254 nm)	HCA, PCA and FA	[26]

Table 1. (Continued)

Sample (compounds)	LC conditions	Chemometrics	Ref.
Detarium microcarpum (phenolic compounds)	Lichrocart C18 (125x4 mm I.D., 5 µm) Gradient elution (1 mL/min): (A) water with 1% phosphoric acid, (B) acetonitrile UV-visible detection (λ 280 and 340 nm)	-	[27]
Hypericum laricifolium (phenolic compounds)	Synergi Hydro-RP 80A (150x4.6 mm I.D., 4 µm) Gradient elution (0.7 mL/min): (A) water with 0.1% trifluoroacetic acid, (B) methanol UV-visible detection (λ 254 nm)	-	[47]
Medicinal plants (phenolic compounds and flavonoids)	XBridge ODS (150x4.6 mm I.D., 5 µm) Gradient elution (1 mL/min): (A) water with 0.5% formic acid, (B) acetonitrile PDA (λ 210 to 500 nm)	-	[29]
Cystophora retroflexa (retroflexanone)	Agilent Eclipse Plus C18 (150x4.6 mm I.D., 5 µm) Gradient elution (1 mL/min): (A) heavy water, (B) acetonitrile NMR detection	-	[30]
Aconitum carmichaelii Debx. (alkaloids and fingerprint analysis)	Agilent Extend C18 (250x4.6 mm I.D., 5 µm) Gradient elution (1 mL/min): (A) water with 0.1% triethylamine, (B) acetonitrile ELSD (carrier gas 1.5 L/min)	HCA, PCA, PLS-DA, OPLS-DA	[31]
Cotton (trace auxins)	Luna C18 (250x4.6 mm I.D., 5 µm) Gradient elution (1 mL/min): water and acetonitrile (50:50 v/v) with 0.1% formic acid FLD detection (λ 230 and 360 nm)	-	[32]
Punica granatum L. flowers (ursolic acid)	Venusil MP C18 (250x4.6 mm I.D., 5 µm) Gradient elution (0.9 mL/min): (A) water with 0.3% formic acid, (B) methanol UV-visible detection (λ 210 nm)	-	[33]
Pyrus communis and *Pyrus pyrifolia* (polyphenols)	Aquasil C18 (250x4.6 mm I.D., 5 µm) Gradient elution (1 mL/min): (A) acetonitrile, (B) 1mM phosphoric acid PDA (λ 190 to 400 nm) FL detection (Ex.: 230, 332, 310, 332 and respectively Em.: 324, 440, 410, 440)	-	[34]

Other stationary phases have also been proposed for the separation of bioactive compounds in natural products. For instance, with the aim to retain polar and hydrophilic analytes in aqueous mobile phase, Chan et al. [44] described the use of a Zorbax SB-Aq reversed-phase alkyl column for the determination of phenolic compounds in six common dietary spices and medicinal herbs. Due to the necessity to achieve a good separation of many compounds in a complex matrix, a gradient elution program of 95 min was employed. In another application, a BEH Phenyl column was proposed to develop a HPLC-UV fingerprinting method for the analysis of *Chimonanthus salicifolius* by Liang et al. [45]. After some studies with other columns the authors found that the BEH Phenyl column was the better option due to its high resolution in the separation and its capacity of retention.

The separation was achieved under gradient elution within 21 minutes. Another study for the simultaneous quantification of rutin and chlorogenic acid in leaves of *Ribes* L. was carried out by Kendir et al. [46] also employing a BEH Phenyl column due to their good selective performance. However, in that case they added 1 mL of triethylamine to the mobile phase constituted by acetonitrile and acidified water. The use of that additive in the mobile phase improved the separation efficiency and allowed to achieve a satisfactory determination of the analytes of interest. A related mobile phase additive, diethylamine, was employed by Lin et al. [37] for the quantitative determination of peimine and peiminine in *Fritillariae Thunbergii Bulbus*. The separation was achieved in 25 minutes and using gradient elution with water with 0.03% diethylamine and acetonitrile as mobile phase components. Again, the use of diethylamine allowed to achieve good chromatographic resolution and baseline, sharp and symmetrical peaks.

An Hydro-RP column was employed by Nimesia et al. [47] for the characterization of phenolic compounds in *Hypericum laricifolium* Juss. The use of that column allowed the separation of bioactive natural compounds in this complex matrix in less than 40 minutes due to their capacity to retain hydrophobic substances. Besides, the use of trifluoroacetic acid in the mobile phase provided stability to the column. Other example that employed gradient elution with trifluoroacetic acid, but in that case using a C18 reversed-phase column, is the work proposed for the quantification of phenolic compounds in *C. fistula* flower extracts by Limtrakul et al. [42]. Figure 1 shows the HPLC-UV fingerprint of a *C. fistula* flower extract with some of the phenolic compounds identified. The HPLC fingerprint was obtained using reversed-phase C18 column and its mobile phase consisted of methanol and 0.1% trifluoroacetic acid, with a flow rate of 1.0 mL/min and the detection wavelength was 280 and 325 nm. Separation was achieved in less than 30 min. The total phenolic content in the analyzed flower extract was 275.32 ± 14.21 mg of gallic acid equivalent (GAE)/g extract. The hydroxybenzoic acid derivatives, including vanillic acid and protocatechuic acid, were found to be 0.95 ± 0.09 and 2.56 ± 0.42 mg/g extract, respectively, whereas gallic acid was recorded at 0.60 ± 0.02 mg/g extract. The hydroxycinnamic acid derivatives, including coumaric acid, ferulic acid, and chlorogenic acid, were detected at 0.39 ± 0.08, 0.80 ± 0.09, and 0.83 ± 0.03 mg/g extract, respectively. The flavonoid content was 27.62 ± 3.56 mg of catechin equivalent (CE)/g extract. Catechin, as a flavonol derivative, was measured at 1.10 ± 0.00 mg/g extract.

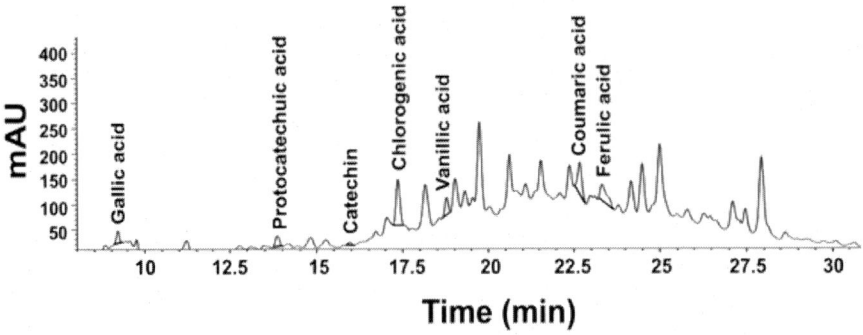

Figure 1. HPLC fingerprint of C. fistula flower extract. The HPLC fingerprint of the flower extract was evaluated using reversed-phase C18 column and its mobile phase consisted of methanol and 0.1% trifluoroacetic acid (TFA). The flow rate was set at 1.0 mL/min and the detection wavelength was 280 and 325 nm. Reprinted with permission from Open Access reference [42].

Another modification in the mobile phase described in the literature was the use of an aqueous 0.2 M $LiClO_4$ in 0.006 M $HClO_4$ and acetonitrile as mobile phase components proposed by Olennikov et al. [24] for the determination of phenolic compounds in *Potentilla anserine* L. by C18 reversed-phase chromatography.

Regarding the detection, liquid chromatography coupled to ultraviolet detection (LC-UV) is the most common method proposed in the literature to carry out the quantitative and qualitative analysis in plant matrices [24–29, 33, 35–47]. For example, Meda et al. [27] described the development of a rapid and effective HPLC-UV method for the simultaneous determination and identification of five phenolic compounds as markers to provide a quality control in the leaves extracts of African tree *Detarium microcarpum* commonly employed in medicine and as food. The separation was performed in a reversed-phase column in a total of 60 minutes, using gradient elution with water acidified with 1% phosphoric acid and acetonitrile. The detection was performed from 200 nm to 400 nm in a L2450 diode array detector (DAD). Another HPLC-UV application was proposed for Megeressa et al. [39] for the study of latex components, in the medicinal plant *Aloe trigonantha L.C.*, to prove their potential in front of a microbes panel. Aloesin, 8-O-methyl-7-hydroxyaloin A/B, aloin A/B, and aloin-6'-O-acetone A/B were the four major compounds identified by UV, as well as with the help of other detection strategies. For example, Aloesin and aloin A/B were also confirmed by mass spectrometry using electrospray in negative ionization mode. Besides, the use of UV detection at 215, 244, 253 and 295 nm permitted to establish the chromone skeleton in Aloesin, together with the help of ^{1}H and ^{13}C RMN spectra. The other two compounds, 8-O-methyl-7-hydroxyaloin A/B, and aloin-6'-O-acetone A/B were confirmed by mass spectrometry with electrospray in positive ionization mode. In addition, the use of RMN spectroscopy and UV at 199, 222, 294 and 348 nm allowed to confirm that 8-O-methyl-7-hydroxyaloin A/B was an hormone derivate while aloin-6'-O-acetone A/B was an hormone rest (UV at 200, 269, 293, 359 nm).

In another example, Singleton et al. [30] employed a HPLC-NMR method to bring about a Mosher analysis to determine the absolute configuration of retoflexanone in *Cystophora retroflexa*. Moshed is a spectroscopy method that allow establishing the absolute configuration of secondary alcohols based in the comparison of the ^{1}H RMN spectra obtained from Moshed prepared esters with those obtained from the original natural product. Figure 2 shows, as an example, the stop-flow HPLC-NMR expansions of the ^{1}H NMR spectra of retroflexanone (1) and Mosher ester derivatives (1a and 1b) showing characteristic chemical shift influences. Characteristic upfield and downfield shifts of the ^{1}H NMR chemical shifts were noted depending on their occurrence on either side of the stereogenic secondary alcohol. These applications demonstrate that the use of NMR detection is sometimes quite decisive when it comes to finding the structure of unknown compounds.

Fluorescence (FL) detection is also very common in combination with LC techniques to study natural products [32, 34]. For instance, to determine four trace auxins (indole acetic acid, indole propionic acid, indole butyric acid and naphthalene acetic acid) in *Cotton*, Zeng et al. [32] employed a LC-FL method coupled to a graphene oxide-cotton surface-solid phase extraction (GO-CF-SPE) procedure. After several studies to study the extraction conditions, graphene oxide was inserted in a cotton surface coupled to a SPE system to achieve auxins extract efficiently and rapidly. Then, the extracts were analyzed by LC-FL method employing an excitation wavelength of 230 nm and an emission wavelength of 360 nm. Besides, the proposed GO-CF-SPE-LD-FL method was validated showing good method performance in terms of limit of detection (LOD), limit quantification (LOQ), precision, repeatability, linearity, and extraction recoveries. In another application, simultaneous UV and FL detection was employed by Sroka et al. [34] for the determination of polyphenols in extracts from leaves of *Pyrus communis* and *Pyrus pyrifolia* collected in their vegetative period. For the FL detection optimized wavelengths of excitation and emission to hydroquinone and hydroxycinnamic that were used as reference compounds were employed. The comparison of the results obtained by UV detection and FL detection showed that fluorescence provided, as expected, better sensitivity and selective in the determination of natural product compounds of interest, allowing the reduction of both LOD and LOQ values.

Trends of Liquid Chromatography, Mass Spectrometry ...   15

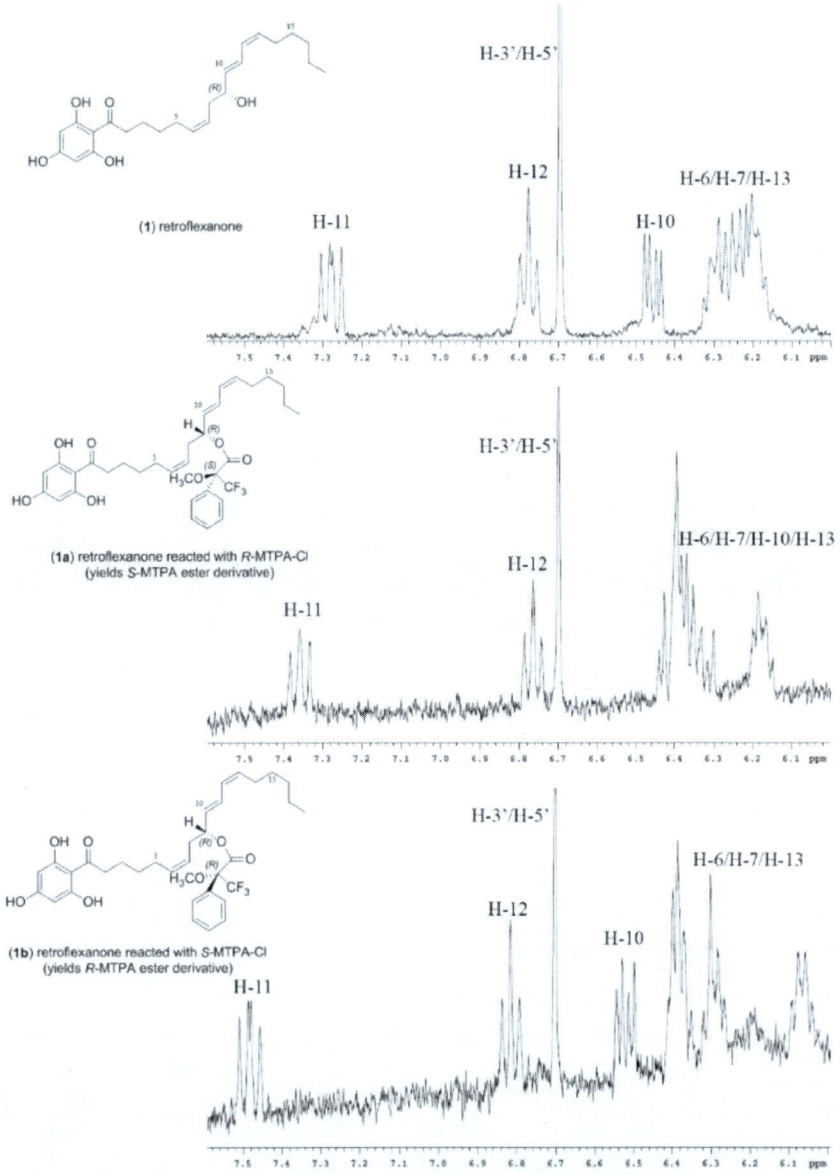

Figure 2. Stop-flow (HPLC-NMR) expansions of the ^{1}H NMR spectra of retroflexanone (1) and Mosher ester derivatives (1a and 1b) showing characteristic chemical shift influences. Reprinted with permission from Open Access reference [30].

Other methods have also been proposed for the detection of bioactive natural components in natural products. For instance, Luo et al. [31] described the use of evaporative light scattering detection (ELSD) for the simultaneous determination of amino alcohol alkaloids and ester alkaloids in the fingerprinting analysis of *Aconitum carmichaelii* Debx. Previous studies had used UV detection for the quantification of these compounds in that matrix, but in the case of amino alcohol alkaloids it was very difficult to obtain good information about them because of the lack of chromophores. Therefore, ELSD was proposed as an alternative detection system for that purpose. ELSD allowed the detection of compounds in non-volatile samples by using a volatile HPLC eluent which is converted in a spray and is heated to evaporate the mobile phase. The dried particles that exit the drift tube are striking with a light that is scattered. The photons are then detected by a photodiode or a photomultiplier. For ELSD the established parameters were a temperature of 50 °C in the evaporation tube, a temperature of 40 °C in the drift tube and a 1.5 L/min flow rate of the carrier gas. The optimization of these parameters, which was done by studying the signal-to-noise ratio, was very critical for the ELSD sensibility and selectivity.

As can be seen in Table 1, regarding sample data treatment most of the proposed methodologies perform only the quantitative determination of the studied targeted bioactive compounds. However, in some of the proposed methodologies employed for the characterization of natural products, especially those performing fingerprinting approaches, a high amount of sample chemical information is provided requiring the use of chemometrics for data analysis [26, 31, 40, 41, 45, 46]. In most of the cases, multivariate analytical methods are proposed. The most frequently chemometric methods employed for the classification of samples are principal component analysis (PCA) and partial least squares regression-discriminant analysis (PLS-DA) [31, 40]. Examples of other chemometric methods such as hierarchical cluster analysis (HCA) [26, 31, 40, 45], similarity analysis (SA) and factor analysis (FA) [26] are also described in the literature. As an example, Luo et al. [31] employed several chemometric methods to classify *Aconitum carmichaelii* Debx. samples from four different regions. HCA data analysis carried out with a IBM SPSS Statistics 23.0 software, was performed based on the contents of nine alkaloids, allowing the perfect classification of the analyzed samples into four groups. PCA, PLS-DA and orthogonal projections to latent structures-discriminant analysis (OPLS-DA), carried out with a SIMCA-P 13.0 software, was performed employing the fingerprints obtained by HPLC-ELSD to address the sample classification and authentication. PCA was used to evaluate the discrimination capacity of all the components. Because the PCA model only provided a preliminary overview of the differences and similarities of the samples of different regions, PLS-DA and OPS-DA models were used to obtain information about which compounds were the most relevant to classify the samples. As expected, PLS-DA provided better discrimination than the one achieved by PCA. In fact, PLS-DA was able to distinguish among the different regions of origin, allowing the authentication of the analyzed plant extracts.

Similarity and factor analysis (SA and FA) chemometric methods are less employed in comparison to the chemometric methods commented in the previous example. Nevertheless, Song et al. [26] employed SA and FA to study the HPLC-UV fingerprints obtained from *Gentiana crassicaulis* extracts. SA tests were performed chosen 21 chromatograms with similar HPLC fingerprints and compared them by Similarity Evaluation System for Chromatographic Fingerprint of Traditional Chinese Medicine (version 2004A). FA was performed by SPSS19.0 software showing that the first seven principal components were relevant to evaluate the quality of the samples.

The chemometric analysis is therefore a good strategy to find discriminant sample chemical descriptors that can be useful in future as biomarkers to authenticate natural products.

LIQUID CHROMATOGRAPHY-MASS SPECTROMETRY TECHNIQUES

In recent years, liquid chromatography coupled with mass spectrometry (LC-MS) methodologies have been used to characterize the chemical composition in complex matrices [48]. In general, target LC-MS methods using triple quadrupole mass analyzers owns advantages such as high robustness, sensitivity and reproducibility [49].

In low resolution mass spectrometry (LRMS), triple quadrupole and ion trap analyzers are the most frequently employed. There are also hybrid variations such as quadrupole in combination with IT (conventional or linear ion traps), i.e., Q-Trap or Q-LIT, among others. Triple quadrupoles are characterized for having unit resolving power and low accuracy. Hoffman et al. [50] described triple quadrupoles resolutions at R = 2000 FHWM (full width at half maximum) (at m/z 1000), accuracy <100 ppm and mass range up to 4000 m/z. However, Ekman et al. [51] mentioned that with hyperbolic rods the transmission is significantly higher and a R = 25000 FHWM (at m/z 1000) is obtained. Ion traps are mainly a modification of quadrupoles, forming a closing loop. Hoffman et al. [50] described resolutions at 4000 FHWM (at m/z 1000), accuracy <100 ppm and mass range up to 6000 m/z. Linear ion traps resolution is improved up to 30000 FHWM (at m/z 520).

Table 2 shows a selection of LC-LRMS methodologies for the analysis and characterization of plant natural products using QqQ (triple quadrupoles) and IT (ion trap) analyzers. As can be seen in the Table, reversed-phase chromatography using C18 columns is frequently employed to address the chromatographic separation, although other stationary phases such as phenyl [52] or C8 [53] are also proposed. Among the used columns, the Acquity BEH Phenyl columns provide selectivity for polyaromatic compounds, low column bleed and long column lifetimes compared to C18 columns [54]. The Acquity BEH C8 columns have sorbents less retentive compared to C18 sorbents [54]. These types of columns also provide low column bleed. For the separation of more polar compounds Atlantis T3 columns are proposed [55]. Atlantis T3 columns contain fully-porous silica-based reversed phase C18 sorbents that retain polar compounds [54]. Another column used to retain polar compounds is the Pack-ODS-AQ column [56, 57], which presents a strong retention in aqueous mobile phases because the C18 ligands lifted from the surface by the action of the eluent that penetrates to the silica pores.

Regarding the mobile phase components, water, acetonitrile and methanol are the solvents most frequently used, usually acidified with formic acid, although acetic acid is also proposed. In some cases, the addition of some electrolytes such as ammonium formate and ammonium acetate for buffer solutions are employed.

Describing some examples, Soufi et al. [55] proposed a LC-MS/MS method with a QqQ mass analyzer using an Atlantis T3 column for the determination of steviol glycosides in *Stevia rebaudiana*. The gradient elution program used water and acetonitrile, both of them acidified with 0.05% trifluoroacetic acid, for the chromatographic separation. Chilling stress is one of the most important environmental factors that limit field crop productivity around the world. This study aims to investigate whether pre-treatment with signaling molecules (salicylic acid, H_2O_2, 6-benzylaminopurine and $CaCl_2$) could induce tolerance of *Stevia rebaudiana* plants to chilling stress. To evaluate the effect of signaling molecules, they looked to morphological and physiological parameters and to the variability of steviol glycosides in *Stevia rebaudiana* plants. Two different mass spectrometric approaches were taken. The first approach was a qualitative LC-HRMS method, and the second approach was a quantitative LC-LRMS method. The second one aimed to quantify steviol glycosides in *Stevia rebaudiana* plants to study if there was variability in the content. The results showed that the content of steviol glycosides varied when pre-treatment with signaling molecules was applied.

Another example is the Pack-ODS-AQ column employed by Tessema et al. [56] for the determination of glucosylceramides in Ethiopian plants (*Lathyrus sativus*, *Brassica carinata* and *Phaseolus vulgaris*) using a LC-MS/MS method with a IT mass analyzer. Gradient elution using water and methanol, both with 0.1% formic acid, was proposed for the chromatographic separation.

Table 2. Selected LC-LRMS methodologies for the analysis and characterization of plant natural products

Sample (compounds)	LC and MS conditions	Chemometrics	Ref.
Spatholobi Caulis (phenolic compounds)	Eclipse plus C18 (50x4.6 mm I.D., 1.8 µm) Gradient elution (0.5 mL/min): (A) water with 0.3% formic acid, (B) methanol ESI (+) QqQ (MRM acquisition mode)	-	[58]
Ding-Zhi-Xiao-Wan (chemical constituents)	Kromasil C18 column (250x4.6 mm I.D., 5.0 µm) Gradient elution (0.5 mL/min): (A) water with 0.1% formic acid, (B) acetonitrile ESI (-) IT (full scan 100-1200 *m/z* for Extraction A and 100-2000 *m/z* for the others)	-	[59]
Buxus papillosa (steroidal alkaloids)	C18 column (55x4.6 mm I.D., 3.0 µm) Gradient elution (0.5 mL/min): (A) water with 0.1% formic acid, (B) acetonitrile with 0.1% formic acid ESI (+) QqQ (MRM acquisition mode)	-	[60]
Alismatis Rhizoma (triterpenoids components)	Quantitative analysis Cortecs UPLC C18 column (100x2.1mm I.D., 1.6 µm) Gradient elution (0.3 mL/min): (A) water, (B) acetonitrile ESI (+) QqQ (MRM acquisition mode)	-	[61]
Yinhua Kanggan Tablet (bioactive substances)	Cortecs UPLC C18 column (50x2.1 mm I.D., 1.6 µm) Gradient elution (0.25 mL/min): (A) acetonitrile, (B) water with 0.1% formic acid and 5% methanol ESI (±) QqQ (MRM acquisition mode)	-	[62]
Shexiang Tongxin Dropping Pill (major constituents)	Quantitative analysis Cortecs C18 (100x2.1 mm I.D., 1.6 µm) Gradient elution (0.25 mL/min): (A) acetonitrile, (B) water with 0.1% formic acid ESI (+) QqQ (MRM acquisition mode)	-	[63]

Sample (compounds)	LC and MS conditions	Chemometrics	Ref.
Gualou Guizhi granules (bioactive substances)	Acquity UPLC Cortest C18 column (100x2.1 mm I.D., 1.6 μm) Gradient elution (0.25 mL/min): (A) water with 0.1% formic acid, (B) acetonitrile with 0.1% formic acid ESI (-) QqQ (MRM acquisition mode)	-	[64]
Qi-Fu-Yin (chemical constituents)	Quantitative analysis Zorbax Extend C18 column (150x2.1 mm I.D., 5.0 μm) Gradient elution (0.5 mL/min): (A) water with 0.1% formic acid, (B) acetonitrile with 0.1% formic acid ESI (±) ; ESI (-) QqQ (MRM acquisition mode) ; Q	-	[48]
Withania coagulans extract (steroidal lactones)	Eclipse XDB C18 column (50x3.0 mm I.D., 1.8 μm) Gradient elution (0.9 mL/min): (A) 60% Milli-Q water with 0.1% formic acid, (B) 40% acetonitrile with 0.1% formic acid ESI (+) QqQ (MRM acquisition mode)	-	[65]
Onobrychis viciifolia (proanthocyanidins)	Acquity UPLC BEH Phenyl column (100x2.1 mm I.D., 1.7 μm) Gradient elution (0.5 mL/min): (A) acetonitrile, (B) water with 0.1% formic acid ESI (-) QqQ (MRM acquisition mode)	-	[52]
Eruca sativa, Eruca vesicaria and Diplotaxis tenuifolia (glucosinolate and flavonol)	Zorbax SB C18 column (100x2.1 mm I.D., μ1.8 m) Gradient elution (0.3 mL/min): (A) ammonium formate (0.1%) and (B)acetonitrile ESI (-) IT	PCA	[66]
Guanjiekang (bioactive substances)	Acquity UPLC C18 column (100x2.1 mm I.D., 1.7 μm) Gradient elution (0.35 mL/min): (A) buffer solution (10 mM ammonium acetate containing 0.1% acetic acid), (B) acetonitrile ESI (+) QqQ (MRM acquisition mode)	-	[67]

Table 2. (Continued)

Sample (compounds)	LC and MS conditions	Chemometrics	Ref.
Dioscoreae herbs (steroidal saponins)	Quantitative analysis Cortecs TM UPLC C18 column (50x2.1 mm I.D., 1.6 µm) Gradient elution (0.4 mL/min): (A) water, (B) acetonitrile ESI (-) QqQ (MRM acquisition mode)	PCA HCA	[68]
Mahuang-Fuzi-Xixin decoction (chemical constituents)	Quantitative analysis Diamonsil C18 (150x2.1 mm I.D., 5.0 µm) Gradient elution (0.3 mL/min): (A) water with 0.1% formic acid, (B) acetonitrile ESI (+) QqQ (MRM acquisition mode)	-	[69]
Fufang Biejia Ruangan Pill (bioactive substances)	Acquity TM, UPLC BEH C18 column (50x2.1 mm I.D., 1.7 µm) Gradient elution (0.25 mL/min): (A) water:acetic acid (100:0.1 v/v), (B) acetonitrile:acetic acid (100:0.1 v/v) ESI (±) QqQ (MRM acquisition mode)	-	[49]
Stevia rebaudiana (steviol glycosides)	Quantitative analysis Atlantis T3 column (150x2.1 mm I.D., 5.0 µm) Gradient elution (0.2 mL/min): (A) water with 0.05% trifluoroacetic acid, (B) CH_3CN with 0.05% trifluoroacetic acid ESI (+) QqQ (MRM acquisition mode)	PCA	[55]
Forsythia suspensa (phillyrin)	DL-C18 column (150x4.6 mm I.D., 5.0 µm) Isocratic elution (0.5 mL/min): Acetonitrile and water with 15 mM formic acid (25:75 v/v) ESI (+) IT (full scan (100-1000 *m/z*), SIM mode)	-	[70]
Zhi-Zi-Da-Huang-Tang, Yin-Chen-Hao-Tang and Da-Huang-Xiao-Shi-Tang (chemical constituents)	Zorbax SB-C18 HT column (50x4.6 mm I.D., 1.8 µm) Gradient elution (0.6 mL/min): (A) water with 0.05% formic acid, (B) acetonitrile ESI (±) IT (full scan 100-1500 *m/z*)	-	[71]

Sample (compounds)	LC and MS conditions	Chemometrics	Ref.
Dazhu Hongjingtian capsule (major constituents)	Quantitative analysis Zorbax SB-C18 column (100x3.0 mm I.D., 1.8 µm) Gradient elution (0.5 mL/min): (A) water with 0.1% formic acid, (B) methanol ESI (-) QqQ (MRM acquisition mode)	-	[72]
Dregea sinensis (pregnane glycosides)	Intersil ODS column (250x4.6 mm I.D., 5.0 µm) Isocratic elution (0.2 mL/min): CH_3CN and H_2O (38:62, 41:59, 43:57 v/v) with 5 mM ammonium acetate ESI (+) IT	-	[73]
Fufang Xialian Capsule (chemical profiling)	Acquity UHPLC BEH C18 column (50x2.1 mm I.D., 1.7 µm) Gradient elution (0.3 mL/min): in negative mode ((A) acetonitrile and (C) water with 0.1% formic acid) and in positive ion mode ((B) methanol, (C) water with 0.1% formic acid) ESI (±) IT (full scan 100-1300 *m/z*)	-	[74]
Lathyrus sativus, Brassica carinata and *Phaseolus vulgaris* (glucosylceramides)	YMC-Pack ODS-AQ column C18 (150×2.0 mm I.D., 3.0 µm) Gradient elution (0.3 mL/min): (A) water with 0.1% formic acid, (B) methanol with 0.1% formic acid APCI (+) IT (full scan 100–2000 *m/z*; MS/MS 200–900 *m/z*)	-	[56]
Grapes (plant growth regulators)	Aquity UPLC BEH C18 (100x 2.1 mm I.D., 1.7 µm) Gradient elution (0.3 mL/min): (A) water with 0.1% formic acid and 1 mM ammonium acetate, (B) methanol with 1 mM ammonium acetate ESI (±) QqQ (MRM acquisition mode)	-	[75]
Achillea coarctata and *Achillea monocephala* (phytochemicals)	Inertsil ODS-4 C18 (100×2.1 mm I.D., 2.0 µm) Gradient elution (0.25 mL/min): (A) water with 0.1% formic acid and 10 mM ammonium formate, (B) acetonitrile ESI (-) QqQ (MRM acquisition mode)	-	[76]

Table 2. (Continued)

Sample (compounds)	LC and MS conditions	Chemometrics	Ref.
Echium plantagineum (pyrrolizidine alkaloids)	Zorbax Eclipse XDB C18 column (150×4.6 mm I.D., 5.0 μm) Gradient elution (0.5 mL/min): (A) water with 0.1% formic acid, (B) acetonitrile ESI (+) QTrap (MRM acquisition mode)	-	[77]
Cucumis sativus (antioxidants)	Zorbax StableBond C18 (50x4.6 mm I.D., 3.5 μm) Gradient elution (0.5 mL/min): (A) water with 0.1% formic acid and 5 mM ammonium formate, (B) methanol ESI (±) QqQ (DMRM acquisition mode)	-	[78]
Hypericum perforatum L., *H. calycinum* L., and *H. confertum* (bioactive substances)	Inertsil ODS-4 C18 (100×2.1 mm I.D., 2.0 μm) Gradient elution (0.25 mL/min): (A) water with 0.1% formic acid and 10 mM ammonium formate, (B) acetonitrile ESI (±) QqQ (MRM acquisition mode)	-	[79]
Ephedra alata (phenolic compounds)	Discovery Bio C18 column (250×4.6 mm I.D., 5.0 μm) Gradient elution (0.5 mL/min): water with 0.2% acetic acid and 5% methanol, (B) water:acetonitrile (50:50 v/v) with 0.2% acetic acid ESI (-) Q (SIM acquisition mode)	-	[80]
Symphytum anatolicum (bioactive substances)	Poroshell 120 EC-C18 column (100×4.6 mm I.D., 2.7 μm) Gradient elution (0.4 mL/min): (A) water with 0.1% formic acid, (B) methanol ESI (±) QqQ (MRM acquisition mode)	-	[81]
Propolis (phenolic and flavonoids compounds)	Gemini C18 (50x2.0 mm I.D., 5.0 μm) Isocratic elution (phosphate 0.8 mL/min): Phosphate buffered saline (pH = 4.5) in water and methanol (40:60 v/v) ESI (+) QqQ-LIT (SIM acquisition mode)	-	[82]

Sample (compounds)	LC and MS conditions	Chemometrics	Ref.
Calycotome spinosa (bioactive substances)	Symmetry Shield C18 (50x4.6 mm I.D., 3.5 µm) Gradient elution (0.5 mL/min): (A) acetonitrile:methanol (5:95 v/v), (B) acetonitrile HESI (±) LIT (full scan 110-2000 *m/z*)		[83]
Hyeonggaeyeongyo-tang (HYT), medical herbs (metabolites for quality assessment)	SunFire C18 analytical column (250x4.6 mm I.D., 5.0 µm) Gradient elution (1.0 mL/min): (A) water with 0.1% formic acid, (B) acetonitrile with 0.1% formic acid ESI (±) QqQ (MRM acquisition mode)	-	[84]
Ziziphus jujuba and *Ziziphus nummularia* (bioactive substances)	Quantitative analysis Zorbax SB-C18 (50x3.0 mm I.D., 1.8 µm) Gradient elution (0.5 mL/min): (A) water with 0.1% formic acid, (B) methanol with 0.1% formic acid ESI (-) IT (MRM acquisition mode)	PCA OPLS-DA	[85]
Spinacia oleracea (lutein)	Acquity RBEH UPLC C8 column (50x2.1 mm I.D., 1.7 µm) Gradient elution (0.5 mL/min): (A) water with 0.1% formic acid, (B) acetonitrile, (C) 2-propanol APCI (+) QqQ (MRM acquisition mode)	-	[53]
Mentha pulegium (polyphenols)	Fortis C18 column (150x3.0 mm I.D., 3.0 µm) Gradient elution (0.3 mL/min): (A) water with 0.1% formic acid, (B) methanol with 0.1% formic acid ESI (-) QqQ (MRM acquisition mode)	-	[86]
Medicinal and aromatic plants (bioactive substances)	Poroshell 120 EC-C18 (150x2.1 mm I.D., 2.7 µm) Gradient elution (0.5 mL/min): (A) water with 0.1% formic acid and 5 mM ammonium formate, (B) methanol with 0.1% formic acid and 5 mM ammonium formate ESI (±) QqQ (MRM acquisition mode)	-	[87]

Table 2. (Continued)

Sample (compounds)	LC and MS conditions	Chemometrics	Ref.
Penthorum chinense (chemical constituents)	XR-ODSII C18 column (100x2.0 mm I.D., 2.2 µm) Gradient elution (0.2 mL/min): (A) water with 0.1% formic acid, (B) acetonitrile with 0.1% formic acid HESI (+) QqQ (MRM acquisition mode)	-	[88]
Convallaria majalis (steroids)	Acquity CSH C18 column (100x2.1 mm I.D., 1.7 µm) Gradient elution (0.4 mL/min): (A) water with 0.1% formic acid, (B) 95% acetonitrile with 0.1% formic acid ESI (+) QqQ (MRM acquisition mode)	-	[89]
Astragalus sieberi (phytochemicals)	Acquity UPLC BEH C18 (50x2.1 mm I.D., 1.7 µm) Gradient elution (0.2 mL/min): (A) water with 0.1% formic acid, (B) methanol with 0.1% formic acid ESI (-) QqQ (full scan 100-1000 m/z)	-	[90]
Cucurbita moschata (polyphenols)	Inertsil ODS-4 C18 column (150x3.0 mm I.D., 3.0 µm) Gradient elution (0.5 mL/min): (A) water:methanol (95:5 v/v) with 0.2% acetic acid, (B) acetonitrile:water (50:50 v/v) with 0.2% acetic acid ESI (-) Q	-	[91]
Astragali Radix (AR), herbal medicine (flavonoids and triterpenes)	Acquity UPLC C18 column (2.1 × 100 mm, 1.7 µm) Gradient elution (0.3 mL/min): (A) 0.1% acetic acid and (B) acetonitrile ESI (+) QqQ (MRM acquisition mode)	PCA HCA	[92]

LRMS acquisition was performed in *full scan* mode (m/z 100-2000) and MS/MS (m/z 200-900). The MS/MS fragmentation experiments were performed at different relative collision energies, between 10 and 100%, to monitor the progress of dissociation. Different skin diseases such as psoriasis or atopic dermatitis, are associated with low levels of skin ceramides. Although many glucosylceramides-based dietary supplements aimed to improve skin barrier, there are limited sources of plant glucosylceramides. Therefore, this study aims to determine the presence of glucosylceramides in the three Ethiopian plants (mentioned above). The results showed that *Lathyrus sativus* and *Phaseolus* vulgaris were potential alternative sources of glucosylceramides.

Wang et al. [67] proposed an UHPLC-MS/MS method with a QqQ mass analyzer using an Acquity UPLC C18 column for the determination of thirteen bioactive substances in *Guanjiekang*. Six different mobile phases were examined, and the optimal one was the mobile phase that showed the lower pressure, the best baseline stability, and the highest ionization efficiency. The mobile phase employed consisted of acetonitrile and a buffer solution of 10 mM ammonium acetate containing 0.1% acetic acid. The establishment of a method to determine bioactive substances in pharmaceutical herbs preparations is necessary to ensure quality, safety and efficiency of herbal products and to establish quality control standards. This study aimed to evaluate the quality of a pharmaceutical herbal preparation, *Guanjiekang*, for treating rheumatoid arthritis. Thirteen compounds contained in *Guanjiekang* were selected as quality control markers. The developed method was focused on the quantification of representative compounds from each herb, which reflected the quality of *Guanjiekang* preparation. The results showed that the method could be used for quantitative assessment and quality control of *Guanjiekang* products.

Regarding the ionization techniques, electrospray ionization is the most employed (see Table 2) due to its simplicity of use, although atmospheric pressure chemical ionization (APCI) is also used in some cases. Usually, in HRMS, qualitative analyses that employ non-targeted methods are performed. This strategy is carried out in both positive and negative ion modes to detect as many compounds as possible, since it is a non-targeted method and therefore it is not known what is going to be detected. In contrast, in LRMS mainly quantitative analyses that employ targeted methods are carried out. Therefore, target methods usually are performed either in positive or negative ion mode, depending on the physicochemical properties of the compound.

For example, Zhang et al. [58] proposed a LC-ESI-MS/MS method with a QqQ mass analyzer in positive ionization mode for the determination of sixteen phenolic compounds (five flavanols, seven isoflavonoids, three flavanones and one chalcone) in *Spatholobi Caulis*. This herbal medicine is used for the treatment of blood deficiency, rheumatalgia and menoxenia. The quantitative analysis results showed that flavanols were the most abundant compounds in *Spatholobi Caulis* and the developed method could be employed for quality control assay.

In another work, Musharraf et al. [60] proposed a LC-ESI-MS/MS method with a QqQ mass analyzer in positive ionization mode for the determination of six steroidal alkaloids in *Buxus papillosa*. Steroidal alkaloids are studied because they can be used for assessing various pharmacological properties of this plant. The results showed good sensitivity for the quantification of steroidal alkaloids within 15 min run time. The short analysis time allows a large number of samples to be analyzed. In contrast, Yilmaz et al. [76] proposed a LC-ESI-MS/MS method with a QqQ mass analyzer in negative ionization mode for the determination of 37 phytochemicals (15 phenolic acids, 17 flavonoids, 3 non-phenolic organic acids, 1 phenolic aldehyde and 1 benzopyrene) in *Achillea* species (*Achillea coarctata* and *Achillea monocephala*). The *Achillea* species have therapeutic applications. Phytochemical content of ethanol and methanol-chloroform extracts of root and aerial parts of *Achillea* species were studied. The results showed that the various parts of the species have quite different biological effects and chemical content. The results also showed that the developed method was feasible for the determination of phytochemicals in Achillea species. In another work, Mighri et al. [80] proposed a HPLC-ESI-MS method using a single quadrupole (Q) mass analyzer in negative ionization mode for the determination of phenolic compounds in Tunisian *Ephedra alata*. *Ephedra alata* is a medicinal plant that is used in the treatment of allergies, asthma, fever, headaches and nasal congestion, among others. The HPLC-ESI-MS method was validated for the analysis of 33 standards (16 phenolic acids, 7 flavonoid glycosides and 10 flavonoids aglycones). 24 phenolic compounds were identified and quantified in *Ephedra alata* extracts. Phenolic acids and flavonoids present in the extracts were identified by comparison of mass spectra and retention times with the 33 corresponding standards.

As mentioned above, in HRMS it is common to use both positive and negative ionization modes to analyze the maximum number of compounds as it is not known what is going to be detected. In contrast, LRMS usually uses only one ionization mode since it is known the compound family that is going to be analyzed. As an example Liu et al. [93] proposed a LC-MS/MS method for the determination of bioactive constituents from Dazhu Hongjingtian capsule. In the qualitative analysis a total of 32 compounds were previously identified by a UPLC-MS/MS method using a Q-TOF mass analyzer in positive and negative ionization mode. In the quantitative analysis 12 compounds were quantified by a UPLC-MS/MS method using a QqQ mass analyzer in negative ionization mode. These 12 compounds were chosen as markers to assess the quality of the Dazhu Hongjingtian capsule. The qualitative and quantitative method developed in this study had importance for quality evaluation of the Dazhu Hongjingtian capsule.

As can be seen in Table 2, in some cases ESI is employed in both positive and negative ionization modes. As an example, Dong et al. [49] proposed a LC-ESI-MS/MS method using a QqQ mass analyzer for the determination of bioactive substances in Fufang Biejia Ruangan Pill. Fufang Biejia Ruangan Pill is an herbal preparation consisting of 11 medicines, including herbal and animal-derived medicines. This herbal preparation improves liver function and alleviates hepatic injury, among others biological activities. The results showed that a total of 27 compounds (organic acids, terpenoids, flavonoids, phenylpropanoids and alkaloids) were identified by comparing their retention times and mass spectra with standards or literature data. The study was carried out in both positive and negative ionization modes due to the different nature of the compounds analyzed. For example, organic acids showed a better response in negative mode because acidic groups can lose electrons. Alkaloids cannot be detected in negative mode due to the existence of an amine group, so positive mode was applied. This method was able to determine bioactive substances in Fufang Biejia Ruangan Pill.

Atmospheric pressure chemical ionization (APCI) has also been employed for the analysis of natural products. For instance, Tessema et al. [56] proposed a LC-MS/MS method for the determination of glucosylceramides in Ethiopian plants (*Lathyrus sativus, Brassica carinata* and *Phaseolus vulgaris*) using APCI in positive ionization mode. Glucosylceramides belong to the sphingolipids family, which are membrane lipids. Electrospray ionization has been the most common ionization technique used for the quantification of sphingolipids. The APCI ionization source provides good results for the quantification of non-polar compounds and there is limited published data on the analysis of plant glucosylceramides using this ionization source, thus being selected by the authors for that purpose. The results showed that the APCI source provided structural information of glucosylceramides and can be used for the determination of this family of compounds.

Regarding low resolution mass analyzers, triple quadrupole instruments are definitely the most frequently used. It is worth mentioning that many authors decide to combine HRMS with LRMS, where TOF analyzers are used for compound identification while QqQ are used for quantification. As an example, Zhao et al. [61] wanted to analyze triterpenes in natural medicines due to the recent interest to know the chemical constituents of these functional formulas. For that purpose, they first identified 22 compounds with HRMS and from those, 14 representative compounds were quantified in LRMS. The benefits were a clear compound identification (with high accuracy, below 5 ppm) and a fast analysis time (only 7 minutes). As an example, Figure 3 shows the UPLC-QqQ-MS chromatograms in positive ion mode of a mixture of standards and herbal sample registered in multiple reaction monitoring (MRM) mode. The method was then validated and will be very useful for further pharmacological analysis. A similar example is reported by Li et al. [48] where the chemical components of a Chinese medicine made by 7 herbs was reported. They found up to 156 compounds using HRMS, by comparing the fragmentation pathways described in the literature. Then the 26 major components were chosen and the method was validated in LRMS with a separation time of 22.5 minutes.

About the acquisition mode in QqQ, MRM by registering selected reaction monitored (SRM) transitions is very often used. In this acquisition mode it is necessary to know the precursor and product ion and the collision energies prior to the analytical run, but this is not a problem in targeted analysis where studies with commercially available standards are frequently possible.

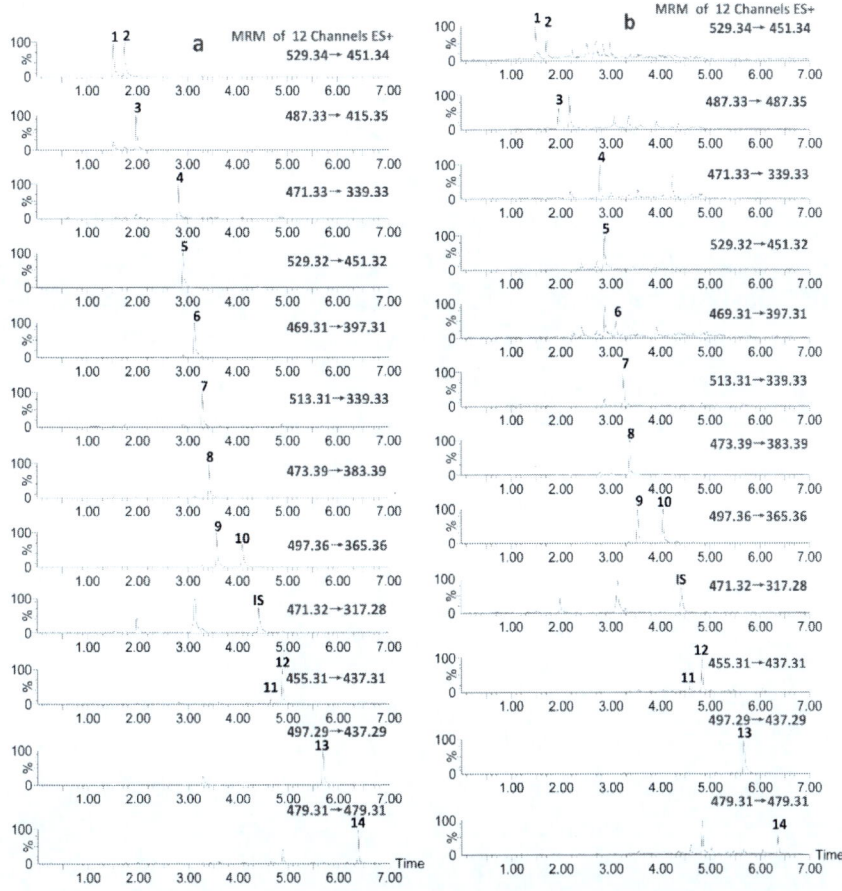

Figure 3. UPLC-QqQ-MS MRM chromatograms in positive ion mode of (a) 14 target standards and (b) herbal sample. Notice fast analysis time in LRMS. Reprinted with permission from Open Access reference [61].

A similar example was reported by Li et al. [48] trying to analyze the components of a new Chinese herb formula. Here, instead of QqQ a LIT was used to determine the structure of the compounds, employing MS^3 fragmentation. Because ion traps have the ability to store ions, it allows to go under multiple MS/MS stages and this allows structure elucidation of unknown compounds.

About the acquisition mode used in IT, *full scan* is frequently employed (the ranges are from 50-2000 Da, depending on the instrument). There are authors as Khan et al. [85] who also propose the use of MRM when using IT type analyzers, although the same sensitivity as in product ion scan will be obtained. In this case, the full potential of the instrument is not exploited, since no significant differences in sensitivity occurred in IT in product ion scan or MRM acquisition modes, while structural information is loss.

As it has been described, the general trend when LRMS is used focus on routine analysis where the goal is to quantify certain phytochemicals. Because of this, only few chemometric methods are reported. The main goal is usually to propose and validate fast methods that can be later used for more complex studies. However, some examples of chemometrics are reported in the literature depending on the application. Among them, PCA is the most common chemometric technique used, and also OPLS-DA has been reported [55, 66, 68, 85, 92].

One example reporting PCA analysis is proposed by Bell et al. [66], which analyzed different glucosinolates and flavonol content of plant salad species (*Diplotaxis tenuifolia, Eruca sativa* and *Eruca vesicaria*) from different commercial varieties. The main goal was the production of new varieties of superior nutritional and sensory quality. Results indicated that there were no significant variances in glucosinolate content but there were in flavonols. As an example, Figure 4 shows PCA scores and loadings plots of PC1 vs. PC2, representing 42% and 14% of the sample variability, respectively. PC1 clearly allows to see two clusters, making a clear differentiation between the two genders *Diplotaxis* and *Eruca*. When we observe the loadings plot, the components that characterize *Eruca* are KDG (kaempferol-3,40-diglucoside) and KG (kaempferol-3-glucoside) while *Diplotaxis* accumulate quercetin and isorhamnetin glucosides in greater amounts. Chemometrics in this study allow to have a better vision of the phytochemical data in different plant genders.

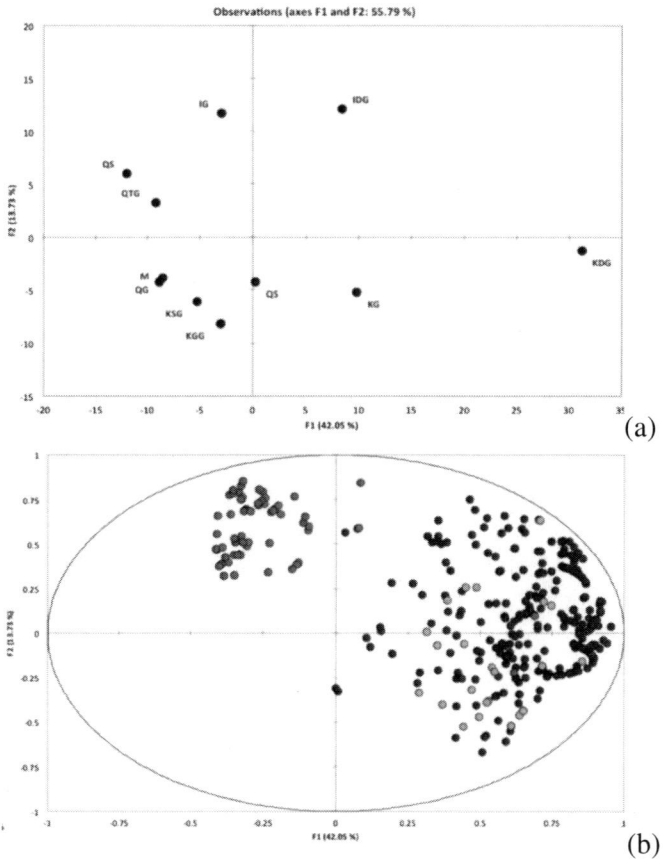

Figure 4. (a) PCA loadings plot of flavonol compounds detected by LC-MS analysis. Abbreviations: M, myricetin; KG, kaempferol-3-glucoside; QG, quercetin-3-glucoside; IG, isorhamnetin-3-glucoside; KDG, kaempferol-3,40-diglucoside; IDG, isorhamnetin-3,40-diglucoside; KGG, kaempferol-3-diglucoside-7-glucoside; QTG, quercetin-3,3,40-triglucoside; KSG, kaempferol-3-(2-sinapoyl-glucoside)-40-glucoside; QC, quercetin-3,40-diglucoside-30-(6-caffeoyl-glucoside); QS, quercetin-3,40-diglucoside-30-(6-sinapoylglucoside). (b) PCA scores plot for individual LC-MS samples tested and their relative distributions in relation to the loadings plot of flavonol composition. Green = *Diplotaxis tenuifolia*; Blue = *Eruca sativa*; Orange = *Eruca vesicaria*. Reprinted with permission from Open Access reference [66].

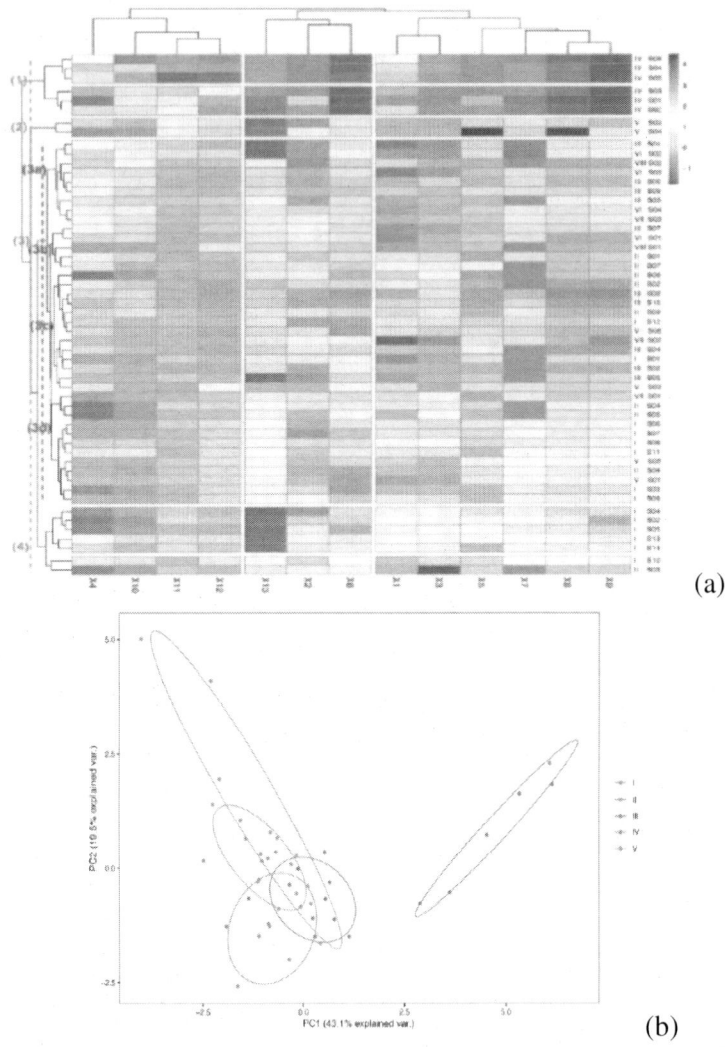

Figure 5. (a) Heatmap presentation of Hierarchical Clustering for 13 compounds. X1, calycosin-7-O-β-D-glucoside; X2, complanatuside; X3, ononin; X4, calycosin; X5, astragaloside IV; X6, astragaloside III; X7, astragaloside II; X8, isoastragaloside II; X9, cyclocephaloside II; X10, formononetin; X11, astragaloside I; 12, isoastragaloside I; 13, β-D-Glucopyranoside, (3β,6α,16β),20R,24S)-3-[(3,4-di-O-acetyl-β-Dxylopyranosyl)oxy]-20, 24-epoxy-16,25-dihydroxy-9,19-cyclolanostan-6-yl. I, Shanxi Province; II, Inner Mongolia; III, Gansu Province; IV, Jilin Province; V, Shaanxi Province; VI, Hebei Province; VII, Northeast China; VIII, Ningxia Province. And (b) PCA scores plot from all 54 samples from different areas I (Shanxi Province), II (Inner Mongolia), III (Gansu Province), IV (Jilin Province) and V (Shaanxi Province). Reprinted with permission from Open Access reference [92].

Another example describing the use of PCA and also hierarchical clustering analysis (HCA) is mentioned by Wang et al. [92] where 14 major compounds (flavonoids and triterpenes) of a Chinese herbal medicine were analyzed. The goal was to use these compounds as chemical markers for quality assessment to identify species (*Astraglus membranaceus* or *Astraglus mongholicus*), growth mode (Cultivated, Semi-wild or Commercial) and production area (different regions of China, Korea and Germany). One of the results, represented in hierarchical clustering, showed the similarities between all samples from the different areas of China (areas I-VIII, forming rows) with different phytochemicals (compounds X1-X13, forming columns) (see Figure 5a).

The results showed four clusters, where cluster 1 consisted of six samples that were all from area IV (Jilin Province), suggesting that the chemical composition distribution in area IV was quite different from the other regions. It was also noticed that components formononetin (11), astragaloside I (12) and isoastragaloside were characteristic of that region. Other clusters (2-4) suggested little variation of the components from the rest of the areas. Then, a PCA scores plot was made in order to confirm results (Figure 5b). The PC1 explained 43.1% of the total variance and PC2 explained 19.5% of the variance. It can be observed that area IV was completely separated from the other regions which confirmed the previous HCA results. Chemometrics in this study helped to understand the phytochemical markers of this Chinese medicinal herb from different regions of China.

LIQUID CHROMATOGRAPHY-HIGH RESOLUTION MASS SPECTROMETRY (LC-HRMS) TECHNIQUES

The number of plant-based naturally occurring secondary metabolites is huge. And the isolation and identification of these potential bioactive compounds is difficult because of their differences in physicochemical properties, structures, and concentration levels, requiring the use of selective and sensitive analytical methodologies. Moreover, most of these compounds that could exhibit health preventing activities may be found at very low concentration levels within a very complex plant-based matrix with thousands of chemicals at higher concentrations. And here is where HRMS techniques, manly using TOF and Orbitrap analyzers, are playing a unique role due to their high resolution power and accurate mass measurements [4, 8, 14, 16]. In general, TOF instruments present a resolving power (instrument's ability to measure the mass of two closely related ions precisely) of approximately 10,000–40,000 full-width at half-maximum (FWHM) with accuracies in the mass determination of 1–5 ppm, while the resolving power of Orbitrap instruments is in the range 10,000–240,000 FWHM with 1–2 ppm mass accuracy (for comparison, a conventional quadrupole MS instrument presents resolving powers of 1,000 FWHM and accuracies of 500 ppm). And this is an important achievement as HRMS can differentiate substances with the same nominal mass-to-charge ratio but different elemental composition (isobaric compounds), providing accurate *m/z* values with four to six decimals. These capabilities also will allow to simplify sample treatment procedures because many interferences (for instance, those matrix chemicals found at higher concentrations) may be removed based on the HRMS accurate mass measurements [8].

However, among plant natural products the number of isomeric compounds is also very high, and it should be taken into account that only some specific isomers may be the ones exhibiting potential bioactive activity. Therefore, fragmentation studies performed in high-resolution MS are also required, and consequently hybrid HRMS instrumentation such as Q-TOF, Q-Orbitrap, and LTQ-Orbitrap, among others, are among the most frequently employed techniques to address the analysis and characterization of plant natural products. It is important not to forget that sometimes the combination of these methodologies with other characterization techniques such as NMR, or the use of the omics strategies, such as metabolomics, in combination with chemometrics, will be necessary to achieve the isolation and identification of new plant natural products [4]. For example, nowadays, the use of ion mobility mass spectrometry (IMS) methodologies, based on the separation of ions in the gas phase because of their different mobility in an inert buffer gas, and consequently, able to separate even isobaric or isomeric compounds [94, 95], are also playing an important role in the characterization, isolation and identification of new plant natural products [96]. However, these techniques will not be addressed in deep in this chapter.

Another important characteristic of HRMS instrumentation is that it maintains very good sensitivity even when working in full scan HRMS acquisition and in a wide m/z range. And the fact that data is consequently being acquired in full scan mode, retrospective analysis of previously registered data without the necessity of running again the samples in the instrument is possible to check if new plant natural products detected and identified in some specific samples were also present in previously analyzed ones.

Tables 3 and 4 show a selection of LC-HRMS methodologies using TOF- and Orbitrap-based analyzers, respectively, for the analysis and characterization of natural products. Regarding LC separations, and following the same trend commented in the previous sections, reversed-phase chromatography using mainly C18 columns are frequently employed, although other stationary phases such as biphenyl ones [97] are also employed. Besides, specially designed C18 reversed-phase stationary phases for the separation of more polar compounds such as the Luna polar C18 column [97], Hypersil Gold aQ RP column [98] or the Acquity UPLC HSS T3 columns, a silica-based bonded phase compatible with 100% aqueous mobile phase [99, 100], have also been proposed. The use of fluorinated stationary phases have also been described [101]. These columns provide complementary separations for many analytes than the ones performed on C18 columns. Different elution orders can be obtained, leading to enhanced selectivity for complicated mixtures of compounds and, in some cases, better separations for highly polar compounds and even isomeric compounds than the ones achieved when using C18 reversed-phase chromatography [102].

Table 3. Selected LC-HRMS methodologies using TOF-based analyzers for the analysis and characterization of plan natural products

Sample (compounds)	LC and HRMS conditions	Chemometrics	Ref.
Tongmai Yangxin Pills (anti-inflammatory compounds)	Zorbax SB-C18 column (250x4.6 mm I.D., 5 µm) Gradient elution (0.6 mL/min): (A) water with 0.05% formic acid, (B) acetonitrile ESI(±) Q-TOF hybrid analyzer (full scan 100 – 1,500 m/z)	PLS	[104]
Radix Scutellariae (antidiabetic constituents)	Luna C18(2) column (150x4.6 mm I.D., 3 µm) Gradient elution (0.5 mL/min): (A) water:acetonitrile (95:5 v/v) with 0.1% formic acid, (B) acetonitrile:water (95:5 v/v) with 0.1% formic acid ESI(+) Q-TOF hybrid analyzer Strategy in combination with LC-HRMS-SPE-NMR	-	[105]
Medicinal cannabis (cannabinoids)	Poroshell 120 C18 column (100x2.1 mm I.D., 2.7 µm) Gradient elution (0.5 mL/min): (A) water with 0.1% formic acid, (B) acetonitrile with 0.1% formic acid ESI(+) Q-TOF hybrid analyzer (full scan 50 – 500 m/z; MS/MS 50 – 1,700 m/z at 20 eV)	-	[106]
Lawsonia inermis L. leaves (Antileishmanial compound profiling)	Luna C18(2) column (150x4.6 mm I.D., 3 µm) Gradient elution (0.5 mL/min): (A) water:acetonitrile (95:5 v/v) with 0.1% formic acid, (B) acetonitrile:water (95:5 v/v) with 0.1% formic acid ESI(±) Q-TOF hybrid analyzer Strategy in combination with LC-HRMS-SPE-NMR	-	[107]
Liang-wai Gan Cao (Radix *Glycyrrhizae uralensis*) (flavonoid components)	DIKMA Leapsil C18 column (150x2.1 mm I.D., 2.7 µm) Gradient elution (0.2 mL/min): (A) water with 0.2% formic acid, (B) acetonitrile ESI(+) Q-TOF hybrid analyzer (full scan 100 – 1,700 m/z)	-	[108]

Sample (compounds)	LC and HRMS conditions	Chemometrics	Ref.
Polygonum orientale (bioactive compounds)	Kinetex C18 column (150x4.6 mm I.D., 5 μm) Gradient elution (0.6 mL/min): (A) water with 0.1% formic acid, (B) acetonitrile with 0.1% formic acid ESI(±) Q-TOF hybrid analyzer (MS/MS 50 – 1,700 *m/z* at 10, 20 and 30 eV)	-	[109]
Coleonema album (coumarins)	Orthogonal separations with: -Luna C18(2) column (150x4.6 mm I.D., 3 μm) -Kinetex PFP column (pentafluorophenyl phase, 150x4.6 mm I.D., 2.6 μm) Gradient elution (0.5 mL/min): (A) water:acetonitrile (95:5 v/v) with 0.1% formic acid, (B) acetonitrile:water (95:5 v/v) with 0.1% formic acid ESI(+) Q-TOF hybrid analyzer Strategy in combination with LC-HRMS-SPE-NMR	-	[101]
Chimonanthus nitens Oliv. Leaf (structural characterization)	Welch C18 column (250x4.6 mm I.D., 5 μm) Gradient elution (0.7 mL/min): (A) water with 0.1% formic acid, (B) acetonitrile ESI(-) Q-TOF hybrid analyzer (MS/MS 100 – 1,250 *m/z* at 35 eV)	PCA OPLS-DA	[110]
Algerian Lichens (bioactive substances)	Acquity BEH C18 column (100x2.1 mm I.D., 1.7 μm) Gradient elution (0.4 mL/min): (A) water with 0.1% formic acid, (B) acetonitrile with 0.1% formic acid ESI(-) Q-TOF hybrid analyzer (full scan 50 – 1,200 *m/z*; MS/MS at collision energies from 10 to 35 eV)	-	[111]
Goldenberry (*Physalis peruviana* L.) calyces (withanolide compounds)	Zorbax Eclipse Plus C18 column (100x2.1 mm I.D., 1.8 μm) Gradient elution (0.5 mL/min): (A) water with ammonium acetate (5mM, pH 3.0 with formic acid), (B) acetonitrile with 0.1% formic acid ESI(+) Q-TOF hybrid analyzer (full scan 50 – 1,100 *m/z*; MS/MS 50 – 800 *m/z* at 110 V)	-	[103]

Table 3. (Continued)

Sample (compounds)	LC and HRMS conditions	Chemometrics	Ref.
Dachuanxiong decoction (bioactive substances)	Zorbax Eclipe Plus C18 column (150x2.1 mm I.D., 1.8 µm) Gradient elution (0.4 mL/min): (A) water with 0.1% acetic acid and 2 mM ammonium acetate, (B) acetonitrile ESI(+) Q-TOF hybrid analyzer (full scan 50 – 1,500 m/z; MS/MS 50 – 1,000 m/z at 10, 20, and 40 eV)	-	[93]
Cecropia species from Panama (bioactive substances)	Kinetex RP C18 column (100x2.1 mm I.D., 2.6 µm) Gradient elution (0.6 mL/min): (A) water with 0.1% formic acid, (B) acetonitrile with 0.1% formic acid ESI(±) Q-TOF hybrid analyzer (full scan 50 – 1,500 m/z; MS^E with collision ramp from 20 to 30 eV)	PCA	[112]
Sugar Maple Bark and Bud hot-water extracts (bioactive substances)	Zorbax SB-C18 column (250x4.6 mm I.D., 5 µm) Gradient elution (0.7 mL/min): (A) water:formic acid (95:5 v/v), (B) acetonitrile:methanol:water (95:5:5 v/v/v) ESI(±) Q-TOF hybrid analyzer (full scan 100 – 1,000 m/z)	-	[113]
Ligusticum chuanxiong hort (bioactive substances)	XAqua C18 column (150x2.1 mm I.D., 5 µm) Gradient elution (0.3 mL/min): (A) water with 0.1% formic acid, (B) acetonitrile with 0.1% formic acid ESI(+) Q-TOF hybrid analyzer (full scan 100 – 1,700 m/z; MS/MS at 15 and 30 eV)	-	[114]
Schisandrae chinensis Fructus (bioactive substances)	Acquity UPLC BEH C18 column (100x2.1 mm I.D., 1.7 µm) Gradient elution (0.3 mL/min): (A) water with 0.1% formic acid, (B) acetonitrile ESI(+) Q-TOF hybrid analyzer (full scan 100 – 1,200 m/z; MS/MS at 6 and 10-40 eV)	-	[115]

Sample (compounds)	LC and HRMS conditions	Chemometrics	Ref.
Qishen granule (bioactive substances)	Acquity UPLC HSS T3 column (100x2.1 mm I.D., 1.8 μm) Gradient elution (0.4 mL/min): (A) water with 0.1% formic acid, (B) acetonitrile with 0.1% formic acid ESI(±) Q-TOF hybrid analyzer (full scan 100 – 1,200 m/z; MS/MS at 10, 20 and 40 eV)	-	[99]
Mountain-cultivated and cultivated Ginsengs (peptides)	Zorbax RRHD Eclipse Plus C18 column (150x3.0 mm I.D., 1.8 μm) Gradient elution (0.4 mL/min): (A) water with 0.1% formic acid, (B) acetonitrile with 0.1% formic acid ESI(+) Q-TOF hybrid analyzer (full scan 50 – 3,000 m/z; MS/MS 50 – 1,700 m/z at 20 eV)	PCA PLS-DA	[116]
Cannabis sativa L. (bioactive substances)	Mediterranean C18 column (250x4.6 mm I.D., 3 μm) Gradient elution (0. mL/min): (A) water:acetonitrile 95:5 (v/v) with 0.1% formic acid, (B) acetonitrile:water 95:5 (v/v) with 0.1% formic acid ESI(+) Q-TOF hybrid analyzer (full scan 60 – 1,100 m/z; MS/MS at 20 and 40 eV)	-	[117]

With respect to the mobile phase components, water, acetonitrile and methanol are the most common solvents employed, frequently acidified mainly with formic acid, although acetic acid is also used [93], and in some cases the addition of other electrolytes such as ammonium formate [11, 93, 103] and ammonium acetate [97], for pH buffering purposes, is also reported.

For example, Capriotti et al. [97] carried out a thorough evaluation of four reversed-phase columns for the untargeted profiling of glucosinolates in cauliflower (*Brassica oleracea* L. var. *botrytis*) by using UHPLC-HRMS with a Q-Orbitrap hybrid mass analyzer. The compared chromatographic separation systems differing in columns (C18 and biphenyl stationary phases) and mobile phase components (see Table 4) were evaluated based on the number of detected and tentatively identified glucosinolates.

Table 4. Selected LC-HRMS methodologies using Orbitrap-based analyzers for the analysis and characterization of plan natural products

Sample (compounds)	LC and HRMS conditions	Chemometrics	Ref.
Fufang herbal medicine decoction (bioactive substances)	Hypersil Gold C18 column (50x2.1 mm I.D., 1.9 μm) Gradient elution (0.3 mL/min): (A) water with 0.1% formic acid, (B) methanol ESI(±) LTQ-Orbitrap hybrid analyzer (full scan 100 – 1,000 m/z; MS^n)	-	[118]
Resin of *Baccharis tola* Phil. From the Atacama Desert (bioactive substances)	Acclaim C18 column (150x4.6 mm I.D., 2.5 μm) Gradient elution (1 mL/min): (A) water with 1% formic acid, (B) acetonitrile ESI(±) Q-Orbitrap hybrid analyzer (full scan 100 – 900 m/z, resolution 70,000 FWHM)	ANOVA	[119]
ShenKang injection (bioactive substances)	Waters BEH C18 column (50x2.1 mm I.D., 1.7 μm) Gradient elution (0.2 mL/min): (A) water with 0.1% formic acid, (B) acetonitrile ESI(±) Q-Orbitrap hybrid analyzer (full scan 80 – 1,200 m/z, resolution 70,000 FWHM; MS with resolution 17,500 FWHM at 20, 30 and 40 NCE)	PCA	[120]
Lippia origanoides (phenolic profile)	Hibar Purospher® STAR RP C18 column (150x3 mm I.D., 3 μm) Gradient elution (0.6 mL/min): (A) water with 0.1% formic acid, (B) methanol with 0.1% formic acid ESI(±) LTQ-Orbitrap hybrid analyzer (full scan 140 – 1,500 m/z, resolution 30,000 FWHM; MS^n at 35 eV)	-	[121]
Ipomoea asarifolia and *Ipomoea muelleri* Plants (Indole diterpenes)	Betasil C18 column (100x2.1 mm I.D., 5 μm) Gradient elution (0.3 mL/min): (A) water with 0.1% formic acid, (B) acetonitrile ESI(+) Orbitrap analyzer (full scan 100 – 800 m/z, resolution 70,000 FWHM)	-	[122]

Sample (compounds)	LC and HRMS conditions	Chemometrics	Ref.
Artemisia capillaris Thunb (Yinchen) (organic acid components)	Agilent TC C18 column (250x4.6 mm I.D., 5 µm) Gradient elution (1 mL/min): (A) water with 0.2% formic acid, (B) acetonitrile ESI(-) LTQ-Orbitrap hybrid analyzer (full scan 100 – 1,200 m/z, resolution 30,000 FWHM)	-	[123]
YiXinShu Tablet (bioactive substances)	Acquity UPLC BEH C18 column (50x2.1 mm I.D., 1.7 µm) Gradient elution (0.2 mL/min): (A) water with 0.1% formic acid, (B) acetonitrile ESI(±) Q-Orbitrap hybrid analyzer (full scan 100 – 1,500 m/z, resolution 70,000 FWHM; dd-MS^2/dd-SIM at 20, 30 and 40 NCE/stepped, resolution 17,500 FWHM)	-	[124]
Angelica Dahurica Radix (bioactive substances)	Waters BEH C18 column (50x2.1 mm I.D., 1.7 µm) Gradient elution (0.3 mL/min): (A) water with 0.1% formic acid, (B) acetonitrile ESI(+) Q-Orbitrap hybrid analyzer (full scan 50 – 800 m/z, resolution 70,000 FWHM; MS/MS at 30 eV, resolution 17,500 FWHM)	PLS	[71]
Plumula nelumbinis ethanol extract (benzylisoquinoline alkaloids)	Hypersil GOLD C18 column (100x2.1 mm I.D., 1.9 µm) Gradient elution (0.2 mL/min): (A) water with 0.1% formic acid, (B) acetonitrile ESI(+) Q-Orbitrap hybrid analyzer (full scan 100 – 1,500 m/z, resolution 35,000 FWHM; ddMS^2 at 35 eV and 25, 35 and 45 NCE/stepped, resolution 17,500 FWHM)	-	[125]
Xiao-ai-ping injection (C_{21} steroids)	Acquity UPLC BEH C18 column (50x2.1 mm I.D., 1.7 µm) Gradient elution (0.2 mL/min): (A) water with 0.1% formic acid, (B) acetonitrile ESI(±) Q-Orbitrap hybrid analyzer (full scan 100 – 1,500 m/z, resolution 70,000 FWHM; MS^2 at 20, 30 and 40 NCE)	-	[126]

Table 4. (Continued)

Sample (compounds)	LC and HRMS conditions	Chemometrics	Ref.
Actinocephalus divaricatus (Eriocaulaceae) (bioactive substances)	Atlantis T3 RP C18 column (150x2.1 mm I.D., 10 μm) Gradient elution (0.2 mL/min): (A) water with 0.1% formic acid, (B) acetonitrile with 0.1% formic acid ESI(-) LTQ-Orbitrap hybrid analyzer (full scan 230 – 1,500 *m/z*, resolution 30,000 FWHM; MS/MS by data-dependent experiments)	-	[127]
Cauliflower (glucosinolates)	Column comparison: - Luna Omega polar C18 column (100x2.1 mm I.D., 1.6 μm), 0.4 mL/min - Kinetex core-shell Biphenyl column (100x2.1 mm I.D., 1.7 μm), 0.4 mL/min - Kinetex core-shell XB C18 column (100x2.1 mm I.D., 2.6 μm), 0.4 mL/min - Two Kinetex core-shell XB C18 columns (100x2.1 mm I.D., 2.6 μm) connected with a zero-dead-volume UHPLC fingertight fitting, 0.6 mL/min Gradient elutions: - (A) water with 0.1% formic acid, (B) acetonitrile with 0.1% formic acid - (A) water with 0.2% formic acid, (B) acetonitrile with 0.2% formic acid - (A) water with 5 mM ammonium formate, (B) acetonitrile with 5 mM ammonium formate - (A) water with 0.1% formic acid, (B) acetonitrile ESI(-) Q-Orbitrap hybrid analyzer (full scan 150 – 1,000 *m/z*, resolution 70,000 FWHM; MS/MS at 40 NCE, resolution 17,500 FWHM)	-	[97]
Dan-Deng-Tong-Nao capsule (bioactive substances)	Acquity UPLC BEH C18 column (50x2.1 mm I.D., 1.7 μm) Gradient elution (0.2 mL/min): (A) water with 0.1% formic acid, (B) acetonitrile ESI(±) Q-Orbitrap hybrid analyzer (full scan 100 – 1,500 *m/z*, resolution 70,000 FWHM; MS/MS at 20, 30 and 40 NCE)	-	[128]

Sample (compounds)	LC and HRMS conditions	Chemometrics	Ref.
Schisandrae chinensis Fructus (bioactive substances)	Acquity UPLC BEH C18 column (100x2.1 mm I.D., 1.7 μm) Gradient elution (0.3 mL/min): (A) water with 0.1% formic acid, (B) acetonitrile ESI(+) LTQ-Orbitrap hybrid analyzer (full scan 100 – 1,200 *m/z*, resolution 30,000 FWHM; MS^n at 35 NCE)	-	[115]
Aechmea magdalenae (Andre) Andre ex Baker, Central American Medicinal Plant (bioactive substances)	Acquity UPLC BEH C18 column (50x2.1 mm I.D., 1.7 μm) Gradient elution (0.7 mL/min): (A) water with 0.1% formic acid, (B) acetonitrile with 0.1% formic acid ESI(+) LTQ-Orbitrap hybrid analyzer (full scan 50 – 500 *m/z*)	-	[129]
Dan-Huang-Qu-Yu capsule (bioactive substances)	Acquity UPLC BEH C18 column (50x2.1 mm I.D., 1.7 μm) Gradient elution (0.2 mL/min): (A) water with 0.1% formic acid, (B) acetonitrile ESI(±) Q-Orbitrap hybrid analyzer (full scan 80 – 1,200 *m/z*, resolution 70,000 FWHM; ddMS^2 at 20, 30 and 35 NCE, resolution 17,500 FWHM)	PCA	[130]
Stachys parviflora L. (bioactive substances)	Kinetex EVO C18 column (5 μm) Gradient elution (0.2 mL/min): (A) water with 0.1% formic acid, (B) acetonitrile with 0.1% formic acid ESI(-) LTQ-Orbitrap hybrid analyzer (full scan 180 – 1,400 *m/z*, MS/MS with data-dependent experiment)	-	[131]
Manchurian Aralia (triterpene saponins)	Hypersil Gold aQ RP column (150x2.1 mm I.D., 3 μm) Gradient elution (0.5 mL/min): (A) water:acetonitrile (95:5 v/v) with 0.1% formic acid, (B) acetonitrile with 0.1% formic acid ESI(±) Q-Orbitrap hybrid analyzer (full scan 100 – 1,000 *m/z*)		[98]

Table 4. (Continued)

Sample (compounds)	LC and HRMS conditions	Chemometrics	Ref.
Centripeda minima L. (sequiterpene lactones)	Acquity UPLC BEH C18 column (50x2.1 mm I.D., 1.7 μm) Gradient elution (0.3 mL/min): (A) water with 0.1% formic acid, (B) acetonitrile:methanol (50:50 v/v) with 0.1% formic acid ESI(+) LTQ-Orbitrap hybrid analyzer (full scan 100 – 1,200 m/z, resolution 120,000 FWHM; MS^2 at 30 NCE, resolution 30,000 FWHM)	-	[132]
Meiguihua oral solution (bioactive substances)	Xbridge BEH Shield RP C18 column (100x2.1 mm I.D., 2.5 μm) Gradient elution (0.2 mL/min): (A) water with 0.1% formic acid, (B) acetonitrile ESI(-) Q-Orbitrap hybrid analyzer (full scan 100 – 1,500 m/z, resolution 70,000 FWHM; MS/MS at 25, 35 and 45 NCE/stepped)	-	[133]
Black currant leaves (phenolic compounds)	Kinetex Evo C18 column (5 μm) Gradient elution (0.2 mL/min): (A) water with 0.1% formic acid, (B) acetonitrile with 0.1% formic acid ESI(-) LTQ-Orbitrap hybrid analyzer (full scan 180 – 1,400 m/z,; MS/MS)	PCA	[134]
Dingkun Dan (bioactive substances)	Acquity UPLC HSS T3 column (100x2.1 mm I.D., 1.8 μm) Gradient elution (0.2 mL/min): (A) water with 0.1% formic acid, (B) methanol with 0.1% formic acid ESI(±) Q-Orbitrap hybrid analyzer (full scan 100 – 1,500 m/z, resolution 70,000 FWHM; MS^2 50 – 1,500 m/z with 25-60 V ramp)	-	[100]
Herbal medicines (antitussive adulterants)	Hypersyl Gold C18 column (100x2.1 mm I.D., 1.9 μm) Gradient elution (0.3 mL/min): (A) water with 0.01 M ammonium acetate, (B) acetonitrile ESI(±) Q-Orbitrap hybrid analyzer (full scan 100 – 1,000 m/z, resolution 70,000 FWHM;	-	[135]

Sample (compounds)	LC and HRMS conditions	Chemometrics	Ref.
	ddMS² at 10-62 NCE, resolution 17,500 FWHM)		
Mountain-cultivated and cultivated Ginsengs (peptides)	Zorbax RRHD Eclipse Plus C18 column (150x3.0 mm I.D., 1.8 μm) Gradient elution (0.4 mL/min): (A) water with 0.1% formic acid, (B) acetonitrile with 0.1% formic acid ESI(+) LTQ-Orbitrap hybrid analyzer (full scan 300 – 2,000 m/z, resolution 60,000 FWHM; MS/MS at 25, 30 and 35 NCE)	PCA PLS-DA	[116]

The authors showed that the Luna polar C18 reversed-phase stationary phase was the one more suitable for the analysis of glucosinolates, due to the relatively high polarity of these compounds, in comparison to Kinetex XB C18 and Kinetex Biphenyl columns stationary phases. Nevertheless, the use of two in series Kinetex XB C18 columns increased the number of tentatively identified glucosinolates in comparison to the use of only one Kinetex XB C18 column, as expected. These results revealed the importance of the column length when dealing with the analysis of complex mixtures such as plant-based matrices. The authors were able to tentatively identify 51 glucosinolates by performing MS/MS experiment when employing the best of the chromatographic systems evaluated. Twenty-four of these compounds were never identified before in cauliflower.

Recently, a polar HSS T3 reversed-phase column was employed by Li et al. [99] for the screening and determination of bioactive constituents in *Qishen* granule by UHPLC-HRMS in a Q-TOF mass analyzer. Gradient elution using water and acetonitrile, both with 0.1% formic acid, was employed for the chromatographic separation. An interesting pseudo-targeted screening strategy based on compound biosynthetic correlations was employed. With this aim, first the authors classified into nine types all possible compounds of *Qishen* granule based on their core skeletons, and the potential analogue molecular formulas were predicted according to core compound-related biosynthetic correlations, such as methylation, hydroxylation, and glucosidation. Then, the authors established several pseudo-compound databases consisting on core compounds, deduced biosynthetic correlations, and predicted analogue molecular formulas. These databases where then employed for screening the obtained sample UHPLC-HRMS raw data, and results were further validated by target tandem mass spectrometry. This strategy allowed the authors to identify 213 constituents in *Qishen* granule, of which 21 were determined as potential new compounds.

The same column was proposed by Gao et al. [100] to perform the chemical profiling of Dingkun Dan, a famous traditional Chinese medicine prescribed for several gynecological diseases, by UHPLC-HRMS using a Q-Orbitrap analyzer in both negative and positive ion modes. In this case, water and methanol, both with 0.1% formic acid, were selected as mobile phase components for the gradient elution chromatographic separation of the analyzed samples. Long elution programs tend to be employed, for instance 45 min in this work, to improve the peak capacity and the chromatographic separation power when dealing with such a complex matrix. HRMS acquisition was performed in full scan mode (m/z 100 – 1,500) with MS/MS fragmentation for top 5 ions to help in the structural identification of the tentatively identified bioactive compounds. A total of 121 components and isomers were characterized by the authors, including amino acids, phenolic acids, lactones, terpenoids, alkaloids, saponins, flavonoids, among others. Besides, a network pharmacology study showed that some of the identified compounds, such as ginsenosides and notoginsenosides, crocin I, echinacoside, rutin and verbascoside, may be related to the Dingkun Dan pharmacological activity.

A pentafluorophenyl column was proposed as an orthogonal separation method following fractionation of a crude ethyl acetate extract of leaves of *Coleoma album* on a preparative-scale C18 column [101], in order to increase the chromatographic separation power for the identification of coumarins by LC-HRMS using a Q-TOF analyzer. More details related to this application will be discussed later.

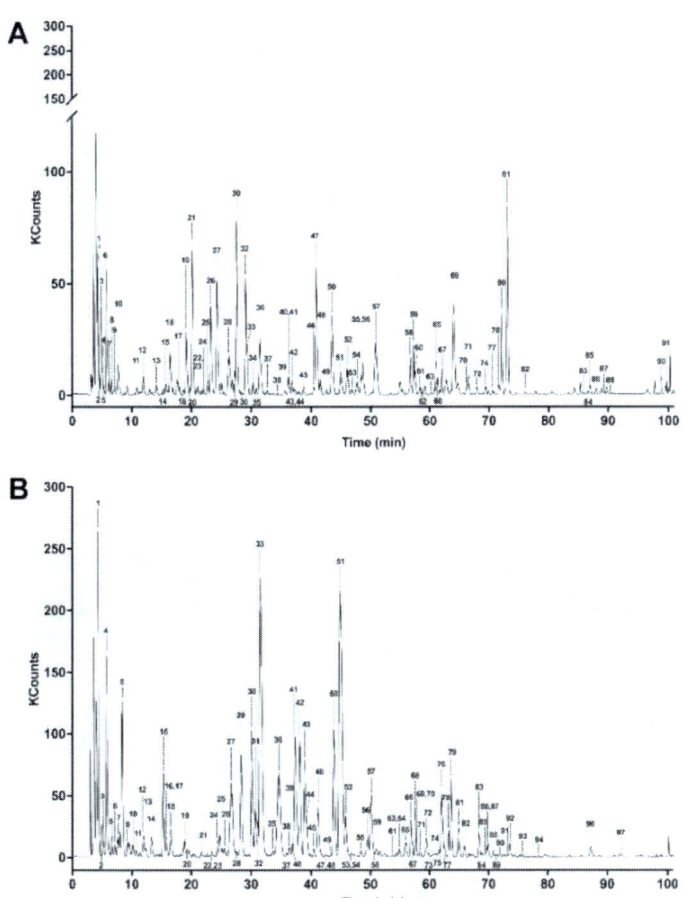

Figure 6. Total ion chromatograms of a sugar maple bud hot-water extract (SBdE) (A) and a sugar maple bark hot-water extract (SBkE) (B) obtained by LC-ESI-QTOF HRMS in negative ion mode. For compound peak identification check the corresponding reference. Reprinted with permission of reference [113]. Copyright (2019) American Chemical Society.

Very important, as previously commented, is the peak capacity and the chromatographic separation power when using LC-HRMS methodologies in full scan HRMS mode for the identification of new natural products in very complex plant material extracts. This is perhaps the reason why although many of the described applications use UHPLC methods (see Tables 3 and 4), normally they are not taking advantage of the sub-2 µm particles or the fused-core with partially porous particles column properties to achieve fast chromatographic separations like in other application fields such as environmental of food safety analysis [136, 137]. Despite the high-resolution and accurate mass measurements achieved by HRMS instrumentation, authors tend to simplify sample extraction treatments in order to reduce discrimination among the extracted compounds, and to increase chromatographic separation time to improve peak capacity detection. That is also the reason why many applications continue employing conventional HPLC columns with 3-5 μm particle sizes, or even higher, like the application described by Zanatta et al. [127], where an Atlantis T3 reversed-phase C18 column (150x2.1 mm I.D., 10 μm particle size) was proposed for a chemical metabolome study associated to the antitumoral activity of *Actinocephalus divaricatus* plant by high-resolution Orbitrap mass spectrometry.

This can be illustrated by Figure 6 showing the metabolite profiling of two maple-derived products, a sugar maple bud hot-water extract (SBdE) and a sugar maple bark hot-water extract (SBkE), using LC-HRMS in a Q-TOF instrument [113].

Chromatographic separation was achieved on a Zorbax SB-C18 column (250x4.6 mm I.D., 5 μm) under gradient elution using water:formic acid (95:5 v/v) and acetonitrile:methanol:water (95:5:5 v/v/v) as mobile phase components. A total gradient elution program of 110 minutes was proposed to increase peak capacity detection and the chromatographic separation of the bioactive metabolites detected. The authors were able to classify almost 100 metabolites in each hot-water extract. SBkE showed to be rich in simple phenolic compounds and phenylpropanoid derivatives, while SBdE contained mainly flavonoids, benzoic acids, and their complex derivatives (condensed and hydrolysable tannins). Approximately 69 phenolic compounds tentatively identified were potentially reported for the first time. Considering the high demand on natural products, these results will help in promoting these two maple-derived products as new sources of bioactive compounds for food, nutraceutical and cosmetic applications.

Electrospray ionization is the most widely employed ionization technique (see Tables 3 and 4), as discussed in the previous section, due to their high versatility and simplicity of use. However, this is expected when used in combination with HRMS for the characterization and identification of plant natural products, as the range of compounds that can be ionized by ESI is high.

This is even more noticeable when screening and non-targeted strategies are employed with LC-HMRS techniques, with the aim of trying to ionize as many compounds as possible, and of different physicochemical properties, without performing any kind of discrimination in the ionization source. Obviously, it should be considered that optimal ionization conditions for all the plant-based extracted compounds is not possible, and great variability on response factors will be observed among the detected compounds. It is also here where the high sensitivity maintained in a wide mass-to-charge ratio range and attainable nowadays by HRMS analyzers, and their resolution and accurate mass measurements, able to discriminate the signal of a given compound from that of interferences typically at higher concentrations, are playing an important role. Besides, not only matrix effect, that with high probability it will be present when dealing with so complex matrices such as plants and plant-based products, is going to affect the ionization efficiency, but it is well known that electrospray ionization suffers of ion suppression/enhancement effects when analyte coelutions are observed [138–140]. This will strongly affect the quantitation of the identified chemical natural products, although in general the qualitative analysis in order to identify new compounds will not be so affected except in the case that the ionization suppression effect is so great that the compound is not detected under the working conditions employed.

Many LC-ESI-HRMS applications dealing with the characterization and identification of new plant natural products are performing electrospray ionization in both positive and negative ion modes (Tables 3 and 4), again with the aim of being able to detect as much compounds as possible (non-targeted strategies), as can be seen in Figure 6. In contrast, some works dealing with specific-compound profiling (target strategies) employ ESI either in positive or in negative ion mode, depending on the physicochemical properties of the analyzed family of compounds. For example, Yu et al. [123] developed a method for the simultaneous quantitation of eight organic acid components (neochlorogenic, chlorogenic, cryptochlorogenic, caffeic,

1,3-dicaffeoylquinic, 3,4-dicaffeoylquinic, 3,5-dicaffeoylquinic, and 4,5-dicaffeoylquinic acids) in *Artemisia capillaris* Thunb (Yinchen) extract by LC-ESI-HRMS using a LTQ-Orbitrap analyzer in negative ionization mode, due to the high ionization efficiency of organic acids in negative ion mode. In another work, D'Urso et al. employed liquid chromatography coupled with high-resolution ESI-LTQ-Orbitrap and ESI-Qtrap MS in negative ion mode for the detection and comparison of phenolic compounds in different extracts of black currant leaves [134]. 31 phenolic compounds mainly belonging to organic acids, flavonoids, catechins and its oligomers were characterized.

In contrast, LC-ESI-HRMS in positive ionization mode was proposed by Lee et al. [122] for the identification of Indole Diterpenes in *Ipomoea asarifolia* and *Ipomoea muelleri*, members of the Convolvulaceae plant family, which are reported to be associated with a tremorgenic syndrome in livestock [141]. Chromatographic separation was performed in a Betasil C18 column under gradient elution using water with 0.1% formic acid and acetonitrile as mobile phase components. As an example, Figure 7 shows the reconstructed LC-HRMS ion chromatograms obtained in positive ion mode for isopropyl alcohol extracts of *I. asarifolia* and *I. muelleri* seeds. HRMS data in combination with MS/MS fragmentation mass spectral data provided valuable information for the indole diterpene characterization. The authors expanded the identification of diterpenes in *I. asarifolia* respect to previous works and reported for the first time their presence in *I. muelleri*. Besides, two new indole diterpenes were isolated and their structures were established by using 1D and 2D NMR spectroscopy (11-hydroxy-12,13-epoxyterpendole K and 6,7-dehydroterpendole A).

Peptides are also among the potential bioactive plant natural products that are typically detected with electrospray ionization in positive ion mode. For instance, recently Zhao et al. proposed the use of peptides as potential biomarkers to address the authentication of mountain-cultivated Ginseng and cultivated ginseng of different ages by employing UHPLC-ESI(+)-HRMS using Q-TOF and LTQ-Orbitrap mass analyzers [116]. Chromatographic separation was carried out in a Zorbax RRHD Eclipse Plus C18 column (150x3mm I.D., 1.8 µm particle size) under gradient elution using water and acetonitrile (both with 0.1% formic acid) as mobile phase components.

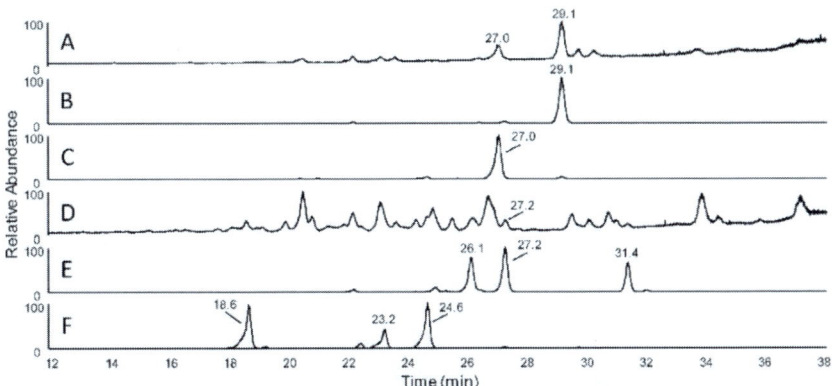

Figure 7. Reconstructed LC-ESI(+)-HRMS ion chromatograms from an isopropyl alcohol extract of *I. asarifolia* seed: (A) total ion chromatogram; (B) *m/z* 518.29009, terpendole K; (C) *m/z* 534.28501, 6,7-dehydroterpendole A, and from an isopropyl alcohol extract of *I. muellleri* seed; (D) total ion chromatogram; (E) *m/z* 518.29009, 11-hydroxy,12,13-epoxyterpendole K and isomers; (F) *m/z* 534.28501, 6,7-dehydroterpendole A and isomers. Reprinted with permission of reference [122]. Copyright (2017) American Chemical Society.

As an example, Figure 8 shows the MS/MS sequencing of the 1172.55 Da peptide named p4 with its spectral identification details. Most of the peaks in the mass spectrum were interpreted automatically by the software; however, a few unsigned peaks needed to be manually appended by the authors to their origins to improve the identification rate. For instance, the y1 ion could only be picked up manually from the spectrum due to the low signal intensity. The authors identified 52 high-confidence peptides and screened 20 characteristic peptides differentially expressed between mountain cultivated and cultivated Ginsengs, some of them expressed significantly only in one type of the samples, that were proposed as robust peptide biomarkers for the discrimination and authentication of Ginseng.

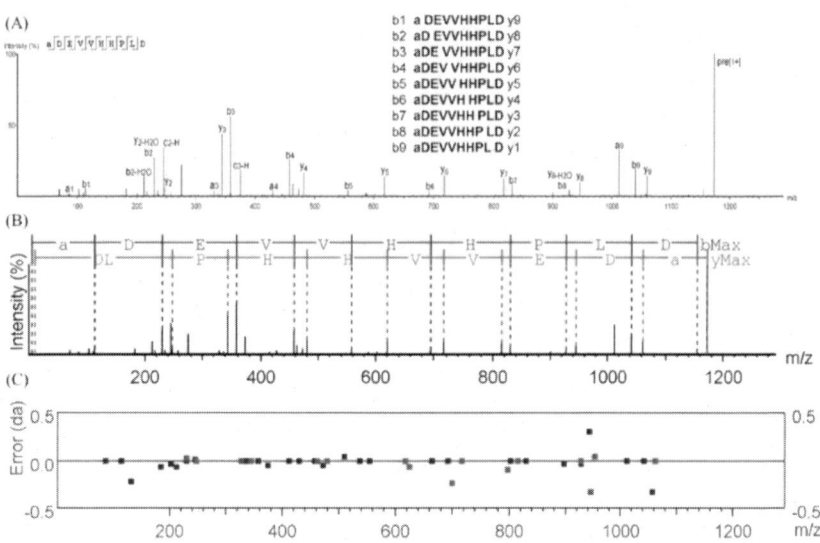

Figure 8. Sequencing of the 1172.55 Da peptide named p4 using database search. (A) MS/MS spectrum; (B) spectrum alignment; (C) error map. Sequence = A (+42.01) DEVVHHPLD; Tag length = 10; −10 log P = 72.73; m/z = 587.2800; z = 2; ppm = −0.6. Reprinted with permission of reference [116]. Copyright (2020) American Chemical Society.

Regarding HRMS analyzers, both time-of-flight (Table 3) and Orbitrap (Table 4) instruments are widely employed in this field. Nevertheless, the use of hybrid instruments such as Q-TOF, Q-Orbitrap and LTQ-Orbitrap are preferred due to the higher instrument capabilities to perform both full scan HRMS and MS/HRMS fragmentation studies, which is required for the correct characterization and structural identification of new plant natural products. One of the main differences among these instruments is the attainable resolution power. While TOF instruments usually permit to work in the range of 10,000 to 40,000 FWHM resolution, Orbitrap instruments exhibit higher resolution power, with commercially available instruments able to attain up to 240,000 FWHM resolution [8], normally at m/z 200, and the value need to be pre-established before performing the analysis. In general, full scan HRMS acquisition tend to be carried out at 70,000 FWHM when Q-Orbitrap instruments are employed, while in the case of LTQ-Orbitrap instruments different resolutions powers are proposed, 30,000 [115, 121, 123, 127], 60,000 [116] or 120,000 [132] FWHM, depending on the application. Obviously this also depends on the pre-established values on the available hybrid instrument.

Regarding MS/HRMS fragmentation studies, resolution need to be decreased when using Orbitrap-based instruments, normally to 17,500 FWHM in Q-Orbitrap analyzers, in comparison to Q-TOF instruments that tend to work in both full-scan at MS/HRMS fragmentation at the same resolution power range due to their higher scan speed capabilities.

Normally, when working with HRMS to address the characterization and identification of plant natural products several acquisition modes are combined to increase the amount of chemical data from the analyzed samples, that will be very useful for the correct identification and confirmation of new bioactive natural product. Therefore, in this field, a high number of acquisition strategies are proposed, depending on the application. Reviewing all these strategies is out of the scope of the present chapter, and only some examples will be illustrated. For instance, Citti et al. evaluated the concentration and the stability of principal cannabinoids in Medicinal cannabis using LC-UV and LC-Q-TOF methods [106]. The acquisition strategy proposed by the authors to evaluate the identity and purity of the determined analytes consisted on a full scan HRMS mode in the scan range 50 – 500 m/z, while MS^2 was automatically performed using nitrogen as the collision gas in the m/z range 50 – 1,700, using the instrument auto MS/MS function and a fixed collision energy of 20 eV. Shin et al. [109] proposed the use of LC-QTOF tandem mass spectrometry in the range 50 – 1,700 m/z at three collision energies (10, 20 and 30 eV) for the characterization and identification of bioactive compounds with potential therapeutic use in *Polygonum orientale*, an annual plant employed as a well-known traditional Chinese medicine.

Similarly, when employing Orbitrap-based analyzers in general full scan HRMS acquisition is combined with different tandem mass spectrometry acquisition strategies. For example, Leitao et al. employed counter-current chromatography in combination with off-line detection by UHPLC-HRMS in a LTQ-Orbitrap hybrid analyzer to study the phenolic profile of *Lippia origanoides* H.B.K. (*Verbenaceae*), a very aromatic shrub or small tree native to some countries of Central and South America [121]. High-Speed

Counter-current Chromatography (HSCCC) was employed for the fractionation of L. origanoides and as a tool for preparative pre-purification to obtain a flavonoid-rich fraction. The authors performed the acquisition in full scan mode within the range 140 – 1,500 *m/z* at a resolution of 30,000 FWHM, in combination with a data dependent scan MS^n fragmentation by keeping the normalized collision energy of the collision-induced dissociation (CID) cell at 35 eV. Compounds were characterized by mass spectra, accurate masses, characteristic fragmentation, and chromatographic retention time. Using this strategy, 12 compounds were further identified by the authors in addition to the major ones previously identified in the non-purified raw extract (i.e., eriodictyol, naringenin and pinocembrin). Two of them, 6,8-di-C-hexosyl-luteolin and 6,8-di-C-flucosyl-apigenin, were reported for the first time in the *Verbenaceae* family.

In contrast, a different tandem mass spectrometry acquisition was proposed by Jiang et al. [125] for the non-targeted bynzylisoquinoline alkaloids characterization from *Plumula nelumbinis* (a traditional Chinese medicine) ethanolic extract by UHPLC-Q-Orbritap HRMS. The authors employed full scan HRMS within the range 100 – 1,500 *m/z*, at a resolution of 35,000 FWHM, in combination with a data dependent fragmentation ($ddMS^2$) by applying stepped normalized collision energies at 25, 35, and 45 eV. This stepped NCE acquisition strategy allow to obtain average MS^2 spectra at the three employed NCEs. In order to process the large volume of data generated by a non-targeted strategy as the one employed in this work, the authors developed an integrated method that combined diagnostic fragment ion filtering (DFIF) with reverse diagnostic fragment loss filtering (RDFLF) to mine non-targeted mass spectral data (check reference [125] for more details on these methods). The use of this strategy enabled at total of 31 benzylisoquinoline alkaloids to be tentatively identified, five of them for the first time.

Fragmentation studies by implying collision energy ramps, able to produce average mass spectral information obtained at low and high collision energies, is also frequently employed for the characterization of plant natural products. This kind of strategy guarantees to obtain enough fragments for identification and confirmation purposes when non-targeted methodologies are employed and it is impossible to know the optimal fragmentation conditions of the detected compounds. For example, Gao et al. [100] evaluated the chemical profile of Dingkun Dan by UHPLC-Q-Orbitrap using full scan HRMS within the range 100 – 1,500 m/z, at a resolution of 70,000 FWHM, in combination with data dependent fragmentation for top 5 ions within the range 50 – 1,500 m/z by using a collision energy ramp of 25-60 V. A total of 121 compounds and isomers were characterized, among them amino acids, phenolic acids, lactones, terpenoids, alkaloids, saponins, and flavonoids.

Very important is the combination of HRMS with other structural characterization methodologies such as NMR. For example, LC-HRMS in combination with solid-phase extraction and NMR (LC-HRMS-SPE-NMR) is a usual practice to address the characterization and identification of new plant natural products [105, 107]. This strategy was employed for the identification of antidiabetic constituents in crude extracts of *Radix Scutellariae* [105]. The authors identified most of the obtained chromatographic peaks from the *Radix Scutellariae* extract as flavonoids or flavonoid glycosides based on the performed HRMS analysis and its comparison with reference data from literature. However, some major components detected were also investigated with NMR in order to confirm the structures and to ascertain structural features which were difficult to conclude from the HRMS data. Several compounds such as ganhuangemin, viscidulin III, baicalin, oroxylin A 7-O-glucuronide, wogonoside, baicalein, wogonin, and skullcap flavone II were then identified. Similar strategy was also employed by Iqbal et al. [107] for the characterization of antileishmanial compounds from *Lawsonia inermis* L. leaves. Active analytes were cumulatively trapped on SPE cartridges and the structures elucidated by NMR spectra obtained in the LC-HRMS-SPE-NMR mode. This allowed the authors the identification of 2,4,6-trihydroxyacetophenone-2-O-β-D-glucopyranoside, lalioside, luteolin-4'-O-β-D-glucopyranoside, apigenin-4'-O-β-D-glucopyranoside, luteolin, and apigenin.

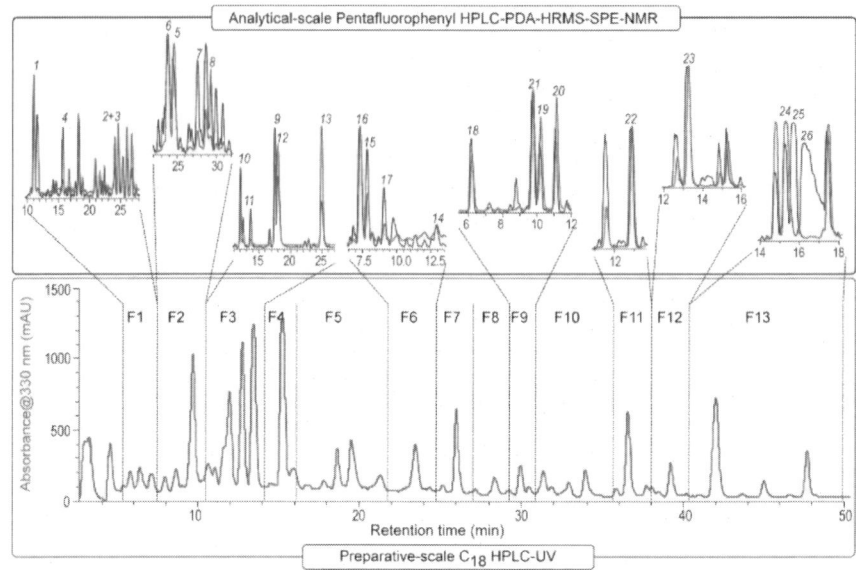

Figure 9. (Bottom) Preparative-scale C18 LC-UV chromatogram at 330 nm of the defatted crude extract of *C. album* leaves. Thirteen fractions were collected as indicated. (Top) Expansions showing base peak chromatograms (blue) and UV chromatograms at 330 nm (red) of selected fractions analyzed by LC-PDA-HRMS-SPE-NMR suing an orthogonal pentafluorophenyl HPLC column. Peak numbering according to elution order on an analytical-scale C18 LC column in a pilot LC-PDA-HRMS-SPE-NMR experiment. Reprinted with permission of reference [101]. Copyright (2017) American Chemical Society.

Most of the LC-HRMS-SPE-NMR investigations reported so far have relied mainly on analytical-scale reversed-phase C18 columns for separation. However, Lima et al. [101] reported the use of an analytical-scale pentafluorophenyl column as an orthogonal separation method following fractionation of a crude ethyl acetate extract of leaves of *Coleonema album*, a shrub that grows along the coastal area of the Cape Peninsula (South Africa), on a preparative C18 column for the determination of coumarins. Obviously, this will allow an important increase in the chromatographic separation power, as can be observed in Figure 9. The proposed setup allowed the LC-PDA-HRMS-SPE-NMR analysis of 23 coumarins, including six new compounds, 8-O-β-D-glucopyranosyloxy-6-(2,3-dihydroxy-3-methylbut-1-yl)-7-methoxycouma-rin, (Z)-6-(4-β-D-glucopyrano-syloxy-3-methylbut-2-en-1-yl)-7-hydroxy-coumarin, 6-(4-β-

D-glucopyranosyloxy-3-methylbut-1-yl)-7-hydroxy-coumarin, (Z)-7-(4-β-D-glucopyranosyloxy-3-methylbut-2-en-1-yloxy) coumarin, (S)-8-(3-chloro-2-hydroxy-3-methylbut-1-yloxy)-7-methoxy-coumarin, and 7-(3-chloro-2-hydroxy-3-methylbut-1-yloxy)coumarin. The separation of several regioisomers that are usually difficult to separate using reversed-phase C18 columns was achieved with the pentafluorophenyl column, showing the potential and wide applicability of LC-PDA-HRMS-SPE-NMR methodologies for accelerated structural identification of plant natural products in complex mixtures when using orthogonal separations.

A very interesting review was recently published by Wolfender et al. [4] showing the trends to improve the metabolite identification in natural product research, going toward the ideal combination of LC-MS/HRMS and NMR profiling, the use of *in Silico* databases, and chemometrics.

As commented in the previous sections, chemometrics is also playing an important role in the characterization and identification of plant natural products, and it is an indispensable tool in many applications dealing with HRMS [4]. Again, bioactive plant natural products may be employed as sample chemical descriptors to study similarities and tendencies among different groups of samples, and PCA, PLS and PLS-DA can be found among the most frequently employed chemometric methods for that purpose. For example, Tao et al. [104] described the use of LC-HRMS for the identification of anti-inflammatory compounds from Tongmai Yangxin Pills, a traditional Chinese medicine, and employed PLS to correlate the relative intensities of the identified compounds with corresponding activities. A total of seven active compounds were then identified according to their coefficient rankings in the obtained PLS model, and their activities were further validated by in vitro bioassay. The results showed that six of these active compounds with high correlation coefficient (R) values exerted certain anti-inflammatory effects in a dose-dependent manner with an anti-inflammatory activity (IC_{50}) values of 53.6 – 204.1 µM.

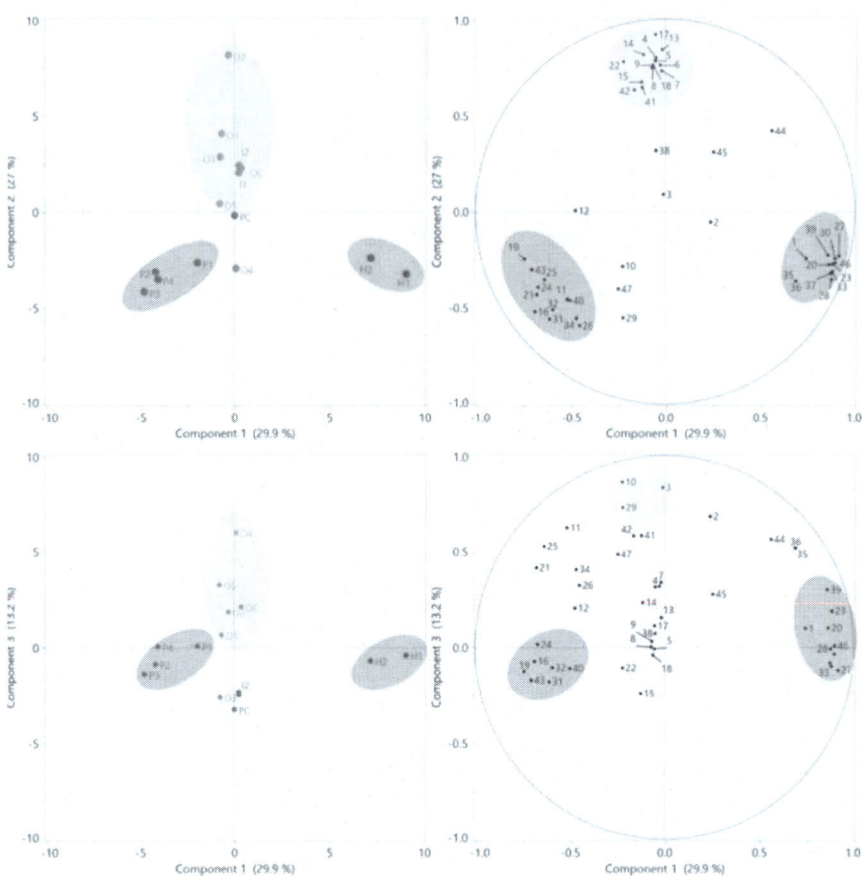

Figure 10. Score plots (left) and loading plots (right) from PCA for *Ceropria* species. (a) PC1 vs. PC2 and (b) PC1 vs. PC3. Reprinted with permission from Open Access reference [112].

Chemometric multivariate data analysis using PCA was, for instance, employed by Rivera-Mondragón et al. for the phytochemical characterization and comparative studies of four *Cecropia* species (C. *obtusifolia*, C. *peltata*, C. *insignis*, and C. *hispidissima*) collected in Panama and typically used as traditional medicines in Latin-America [112]. The authors elucidated the structure of 11 compounds isolated from leaves of C. *obtusifolia* based on HRMS and NMR spectroscopy analysis. Besides, the polyphenolic constituents of leaves of all four *Cecropia* species and several commercial products were characterized using LC-diode-array detection-Q-

TOF, and forty-seven compounds belonging to different families, phenolic acids, flavonoids, flavonolignans, and saponins were tentatively identified. Of those, 31 were not previously reported in the analyzed samples. As an example, Figure 10 shows the PCA score and loading plots (PC1 vs PC2 and PC1 vs PC3) by employing the characterized 47 compounds as chemical descriptors of the analyzed samples.

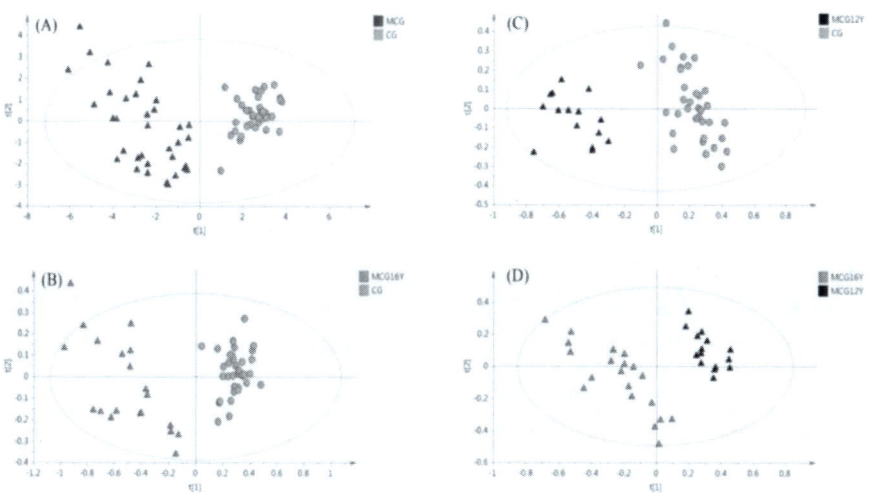

Figure 11. The PLS-DA score plots of P. *ginseng* samples. (A) 16-year-old MCG vs. CG, (B) 16-year-old MCG vs. CG, (C) 12-year-old MCG vs. CG and (D) 16-year-old MCG vs. 12-year-old MCG. Reprinted with permission of reference [116]. Copyright (2020) American Chemical Society.

Both score plots separated the species into three groups. C. *hispidissima* and C. *peltata* individuals were clustered to their respective species, while C. *obtusifolia* and C. *insignis* were mixed. PC1 separated C. *peltata* (showing PC1 negative scores) from C. *hispidissima* (showing PC1 positive scores), while PC2 and PC3 separated C. *obtusifolia* and C. *insignis* (depicting PC2 and PC3 positive scores) from the other species (with PC2 and PC3 negative scores). Considering the importance of the described novel chemical constituents and the increasing interest and use of these natural products, this study will may be of great help for chemotaxonomic purposes, for the interpretation of medicinal properties, and for quality assessment of herbal supplements containing *Cecropia* leaves.

Partial least square regression-discriminant analysis was proposed, for instance, for the authentication of mountain cultivated ginseng (MCG) and cultivated ginseng (CG) of different ages by employing peptides as potential biomarkers of the analyzed samples [116]. As previously described in this section, peptides were determined by UHPLC-ESI(+)-HRMS using both Q-TOF and Q-Orbitrap analyzers. The authors employed PLS-DA analysis for the authentication of MCG and its age, and Figure 11 shows the obtained PLS-DA plots.

For that purpose, samples were divided into 16-year-old and 12-year-old MCG groups to systematically compare the older MCG or younger MCG with the CG samples. The PLS-DA score plots showed that each group was assembled in a specific quadrant of the plot in each PLS-DA model, indicating that there was a significant difference in the composition or content of peptides in these four groups. Therefore, the classification results showed a robust difference of ginseng peptides, indicating the enormous potential of multivariate analysis for MCG authentication.

Two-Dimensional LC Coupled with Mass Spectrometry

Many of the previously commented applications have shown the difficulties in the characterization and identification of plant natural products mainly due to the matrix complexity, and how improving peak chromatographic capacity is required. For this reason, two-dimensional liquid chromatography (2DLC) techniques, where two chromatographic separation phases are combined, might be a good option to increase the separation capacity, and have also been proposed for the characterization and identification of plant natural products [20,96,142–145]. In 2DLC, two main approaches are typically employed: (i) the "heart-cutting" techniques, which is normally abbreviated as LC–LC, and where only some selected fractions coming from the first chromatographic column (first dimension) are transferred to the second chromatographic column (second dimension), and (ii) "comprehensive LC", which is abbreviated as LC × LC, where the entire analyzed sample is subjected to separation in both chromatographic columns (in both dimensions) [96, 146].

For example, comprehensive two-dimensional liquid chromatography (LC × LC) coupled to HRMS by combining hydrophilic interaction chromatography (HILIC) and C18 reversed-phase LC (RP-LC) have shown to be an attractive methodology for the determination of phenolic compounds due to the orthogonal group-type separations provided. However, comprehensive hyphenation of both HILIC and RP-LC is not an easy task due to the very different elution strengths of the mobile phases used in both dimensions, and different approaches have been proposed. As an example, Muller et al. [144] proposed a HILIC×RP-LC-DAD-HRMS methodology for the determination of phenolic compounds by employing dilution of the first dimension flow, and then a large volume injection in the second dimension is derived by kinetic optimization of experimental parameters in order to provide maximum performance. Using this approach, excellent chromatographic performance was described, and a total of 149 phenolic compounds (flavonoid and non-flavonoid phenolics) were tentatively identified in rooibos tea (32 compounds), wine (45 compounds) and grape (72 compounds) samples based on retention data in both dimensions, UV-vis spectral and high- and low collision energy HRMS data, demonstrating the great applicability of 2DLC for the details screening of phenolic components in natural products.

Obviously, 2DLC is also combined with other characterization analytical techniques to enhanced the identification capabilities of the proposed strategy in the analysis of plant natural products. This is the approach employed by Yao et al. [145] for the selective identification of flavonoid O-glycosides from *Carthamus tinctorius*, a traditional Chinese medicine for treating stasis resultant cardiovascular and cerebrovascular diseases. This enhanced targeted identification strategy was achieved by integrating off-line two-dimensional liquid chromatography/LTQ-Orbitrap HRMS, high-resolution diagnostic produc ions/neutral loss filtering and LC-

SPE-NMR spectroscopy. Off-line 2DLC was carried out with an Acchrom XAmide column and a BEH Shield RP-C18 UHPLC column to achieve better separation of the co-eluting components. With this approach, five aglycone structures (kaempferol, 6-hydroxykaempferol, 6-methoxykaempferol, carthamidin, and isocarthamidin) were identified based on the NMR data. Of the 107 characterized flavonoids, 80 flavonoid O-glycosides were reported for the first time in *Carthamus tinctorius*. The authors concluded that this integral strategy can improve the potency, efficiency, and accuracy in the detection of new compounds from medicinal herbs and other natural sources.

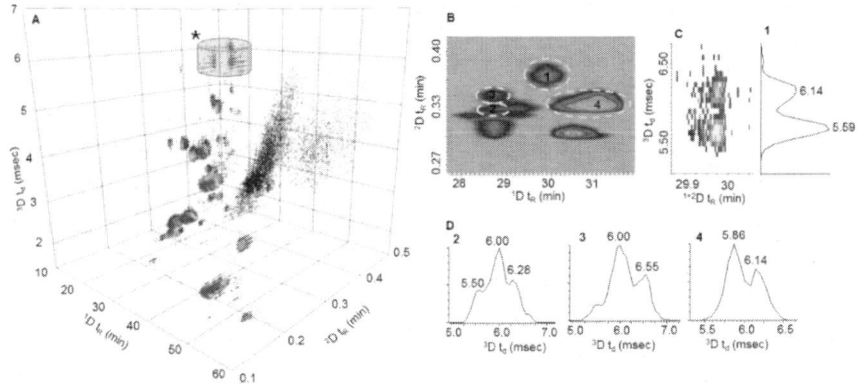

Figure 12. Three-dimensional separation (HILIC x RP-LC x IMS-HRMS) of the procyanidin trimers (*m/z* 865) found in grape seed. (A) Three-dimensional representation of the data. (B) Extracted-ion contour plot for *m/z* 865. (C,D) Mobilograms obtained for the peaks labeled 1-4 on the extracted-ion contour plot. The asterisk indicates the three-dimensional separation space occupied by the procyanidin trimeric species. Reprinted with permission of reference [143]. Copyright (2018) American Chemical Society.

As previously commented in the introduction section, the addition of ion mobility spectrometry (IMS) to LC-MS workflows may offer a means of improving their performance for complex sample analysis by improving the separation of isobaric and isomeric compounds. This separation power is even higher when combined with 2DLC separations and HRMS. As an example, Venter et al. [143] evaluated the potential of combining comprehensive two-dimensional liquid chromatography (LC×LC) and ion mobility spectrometry (IMS)-HRMS. For that purpose, HILIC×RP-LC×IMS-HRMS (Q-TOF analyzer) was used to analyze a range of phenolic compounds of different families (phenolic acids, flavonoids, and hydrolysable and condensed tannins) in several natural products such as a commercially available chestnut tannin extract, red wine, grape seeds, and rooibos tea sample based on a fermented commercial sample of *Aspalathus linearis*. As an example, Figure 12 shows the data obtained by the authors for the HILIC×RP-LC-UV×IMS−HRMS analysis of a grape seed sample, with the data for the procyanidin trimers highlighted in the insets (Figure 12B−D). From the relevant part of the extracted-ion contour plot in Figure 12B, four procyanidin trimers can be observed, which were relatively well-separated because of the combination of HILIC and RP-LC selectivities. However, the extracted-ion mobilograms for each of these peaks (Figure 12C,D) show multiple peak maxima in each case, being most evident for the peak labeled 1. Each of these sets of peaks observed in the drift dimension coeluted following HILIC×RP-LC separation and could not be distinguished based on HRMS data, being only distinguished by IMS. This example highlights the importance of IMS to increase the separation power in very complex matrices.

In another work, a multidimensional analytical approach based on time-decoupling on-line comprehensive 2DLC coupled with ion mobility Q-TOF was proposed for the determination of ginsenosides in white and red ginsengs [142]. Like the previous commented work, this approach enabled the separation of complex matrices in "four" dimensions (2DLC, IMS, and HRMS), enhancing the identification capability. With the proposed methodology, 201 ginsenosides were detected from white and red ginseng samples. Among them, 10 pairs of co-eluting isobaric ginseng saponins that were not resolved by 2DLC-HRMS were further resolved using 2D-UHPLC-IMS-HRMS. Besides, the authors combined the established methodology with unsupervised PCA and supervised orthogonal projections to latent structures-discriminant analysis (OPLS-DA) to discover potential biomarkers able to differentiate red and white ginsengs. Nine of the identified ginsenosides were found to be potential biomarkers for that purpose.

SUMMARY AND CONCLUSION

The importance and the need in research to discover new plant-based natural products is indisputable, due to the possible benefits of these compounds in the prevention of many diseases. For this reason, the number of publications dealing with the characterization, identification and determination of new bioactive natural products is increasing exponentially. Separation analytical techniques are typically employed due to the complexity of plant-based matrices and the variability of chemicals that may be present. In addition, sensitive methodologies are also required because many of these new potentially bioactive chemicals may be found at very low concentration levels. In this chapter, trends of liquid chromatographic techniques, mass spectrometry and chemometrics for the analysis and characterization of plant-based natural products has been reviewed.

The high chromatographic resolution capacity attainable today by liquid chromatography, especially when working under UHPLC conditions, make this technique the method of choice for natural food product characterization. In general, reversed-phase chromatography, mainly using C18 columns under gradient elution conditions with water and methanol or acetonitrile as mobile phase components (sometimes acidified with formic acid), is typically employed because of the grate versatility of this separation mode for a huge range of chemicals. Depending on the application and the physicochemical properties of the families of compounds to be analyzed, other separation conditions based on different stationary phases, such as C8, phenyl stationary phases, or those especially designed for more polar compounds such as Hydro-RP or Atlantis T3 columns, are also employed. Besides, different modifiers are also added to the mobile phase components, i.e., ammonium formate, diethylamine, trifluoroacetic acid, among others, depending on the application. Methods employing liquid chromatography with ultraviolet, fluorescence, NMR and evaporative light scattering detection have been reported for the analysis and characterization of plant-based natural products and some representative examples have been described in this contribution.

Liquid chromatography in combination with both low- and high-resolution mass spectrometry is nowadays one of the best options to address the characterization and identification of new natural products in these complex matrices. In general, compound ionization is achieved by employing electrospray due to its simplicity of use and high versatility, although for some specific applications other ionization sources such as APCI are also reported. Most of the publications employing LC-LRMS methodologies are performed using triple quadrupole instruments, mainly due to the sensitivity and selectivity attainable with these analyzers when working in MRM mode. Besides, most of these applications are developed for quantitative purposes by performing targeted methodologies. Other LRMS instruments, such as those based on ion trap analyzers, or some hybrid combinations, are normally employed when structural information is required for compound identification. However, when addressing the analysis and identification of new unknown bioactive natural products, high-resolution mass spectrometry, either using TOF or Orbitrap analyzers, are the main techniques employed. This is due to the high resolution and the accurate mass measurements capabilities attainable by these spectrometers. The employment of hybrid configurations, i.e., Q-TOF or Q-Orbitrap, is recommended for their fragmentation capabilities under high-resolution conditions, providing additional identification and confirmation criteria to identify new unknown natural products even when commercially standards are not available. Relevant publication examples using both LC-LRMS and LC-HRMS methods for the analysis and characterization of plant-based natural products have also been addressed in the present contribution.

The number of chemical data provided by these methodologies, especially those employing HRMS, is huge, making necessary the use of chemometric methodologies to extract relevant sample chemical features to address the characterization of plant-based natural products. Among the most typically employed chemometric procedures we can find PCA, PLS, PLS-DA, OPLS-DA, HCA, etc., and some examples have also been illustrated in the present contribution.

Finally, multidimensional chromatography in combination with mass spectrometry is also appearing as a powerful strategy to address the analysis and characterization of natural products, mainly due to the high separation capacity achieved when employing orthogonal chromatographic separation modes, in combination with different detection capabilities. Some examples employing even "four dimensions" such as 2DLC, IMS, and HRMS, obviously enhancing the identification capability, have been reported in the literature.

In conclusion, research in plant-based natural products is nowadays a "hot" topic, and the number of publications using liquid chromatography, mass spectrometry and chemometrics will be increasing in the near future.

ACKNOWLEDGMENTS

This work was supported by the Agency for Administration of University and Research Grants (Generalitat de Catalunya, Spain) under the project 2017SGR-310.

REFERENCES

[1] Bhat, S.V., B.A. Nagasampagi, M. Sivakumar, *Chemistry of natural product.*, Springer, New York, USA, 2005.

[2] Hanson, J.R., *Natural Products: the Secondary Metabolites,* The Royal Society of Chemistry, Cambridge, UK, 2003.

[3] Nwokeji, A.P., I.O. Enodiana, E.S. Raymond, O.-I. Osasere, A.H. Abiola, The Chemistry Of Natural Product: Plant Secondary Metabolites, *Int. J. Technol. Enhanc. Emerg. Eng. Res.* 4 (2016) 1.

[4] Wolfender, J.L., J.M. Nuzillard, J.J.J. Van Der Hooft, J.H. Renault, S. Bertrand, Accelerating Metabolite Identification in Natural Product Research: Toward an Ideal Combination of Liquid Chromatography-High-Resolution Tandem Mass Spectrometry and NMR Profiling, in Silico Databases, and Chemometrics, *Anal. Chem.* 91 (2019) 704–742. doi: 10.1021/acs.analchem.8b05112.

[5] Grotewold, E., Plant metabolic diversity: A regulatory perspective, *Trends Plant Sci.* 10 (2005) 57–62. doi: 10.1016/j.tplants.2004.12.009.

[6] Pichersky, E., E. Lewinsohn, Convergent Evolution in Plant Specialized Metabolism, *Annu. Rev. Plant Biol.* 62 (2011) 549–566. doi: 10.1146/annurev-arplant-042110-103814.

[7] Hill, A.F., *Economic Botany: A Textbook of Useful Plants and Plant Products*, McGraw-Hill, New York, USA, 1952.

[8] Aydoğan, C., Recent advances and applications in LC-HRMS for food and plant natural products: a critical review, *Anal. Bioanal. Chem.* (2020). doi: 10.1007/s00216-019-02328-6.

[9] Taiz, L., E. Zeiger, *Plant Physiology*, 3rd ed., Sinauer Associates, California, USA, 2002.

[10] La Barbera, G., A.L. Capriotti, C. Cavaliere, C.M. Montone, S. Piovesana, R. Samperi, R. Zenezini Chiozzi, A. Laganà, Liquid chromatography-high resolution mass spectrometry for the analysis of phytochemicals in vegetal-derived food and beverages, *Food Res. Int.* 100 (2017) 28–52. doi: 10.1016/j.foodres.2017.07.080.

[11] Guo, J., H. Lin, J. Wang, Y. Lin, T. Zhang, Z. Jiang, Recent advances in bio-affinity chromatography for screening bioactive compounds from natural products, *J. Pharm. Biomed. Anal.* 165 (2019) 182–197. doi: 10.1016/j.jpba.2018.12.009.

[12] Mishra, B.B., V.K. Tiwari, Natural products in drug discovery: Clinical evaluations and investigations, In: *Opportunity, challenge and scope of natural products in medicinal chemistry*, Research signpost, Kerala, India, 2011.

[13] Barbosa, S., N. Pardo-Mates, L. Puignou, O. Núñez, The Role of Polyphenols and Polyphenolic Fingerprinting Profiles in the Characterization and Authentication of Natural Food Products, in: J.C. Taylor (Ed.), *Advances in Chemistry Research*. Vol. 40., Nova Science Publishers, Inc., New York, USA, 2017.

[14] Ibekwe, N., S. Ameh, Hyphenated Techniques in Liquid Chromatography as Current Trends in Natural Products Analysis, *Int.*

Res. J. Pure Appl. Chem. 7 (2015) 132–149. doi: 10.9734/irjpac/2015/14458.

[15] Stylos, E., M.V. Chatziathanasiadou, A. Syriopoulou, A.G. Tzakos, Liquid chromatography coupled with tandem mass spectrometry (LC–MS/MS) based bioavailability determination of the major classes of phytochemicals, *J. Chromatogr. B Anal. Technol. Biomed. Life Sci.* 1047 (2017) 15–38. doi: 10.1016/j.jchromb.2016.12.022.

[16] Lucci, P., J. Saurina, O. Núñez, Trends in LC-MS and LC-HRMS analysis and characterization of polyphenols in food, *TrAC Trends Anal. Chem.* 88 (2017) 1–24. doi: 10.1016/j.trac.2016.12.006.

[17] Stavrianidi, A.N., T.M. Baygildiev, E.A. Stekolshchikova, O.A. Shpigun, I.A. Rodin, New Approaches to the Determination and Group Identification of Physiologically Active Compounds in Plant Materials and Commercial Products by High-Performance Liquid Chromatography–Mass Spectrometry, *J. Anal. Chem.* 74 (2019) 58–70. doi: 10.1134/S1061934819010106.

[18] Marshall, A.G., C.L. Hendrickson, High-Resolution Mass Spectrometers, *Annu. Rev. Anal. Chem.* 1 (2008) 579–599. doi: 10.1146/annurev.anchem.1.031207.112945.

[19] Makarov, A., M. Scigelova, Coupling liquid chromatography to Orbitrap mass spectrometry, *J. Chromatogr. A.* 1217 (2010) 3938–3945. doi: 10.1016/j.chroma.2010.02.022.

[20] Li, K.Z., Chen, M.Z. Guo, D.Q. Tang, Two-dimensional liquid chromatography and its application in traditional Chinese medicine analysis and metabonomic investigation, *J. Sep. Sci.* 39 (2016) 21–37. doi: 10.1002/jssc.201500634.

[21] Ji, S., S. Wang, H. Xu, Z. Su, D. Tang, X. Qiao, M. Ye, The application of on-line two-dimensional liquid chromatography (2DLC) in the chemical analysis of herbal medicines, *J. Pharm. Biomed. Anal.* 160 (2018) 301–313. doi: 10.1016/j.jpba.2018.08.014.

[22] Massart, D.L., B.G.M. Vandeginste, L.M.C. Buydens, S. de Jong, P.J. Lewi, J. Smeyers-Verbeke, Handbook of chemometrics and qualimetrics., Elsevier, Amsterdam, The Netherlands, 1997.

[23] Kumar, N., A. Bansal, G.S. Sarma, R.K. Rawal, Chemometrics tools used in analytical chemistry: An overview, *Talanta.* 123 (2014) 186–199. doi: 10.1016/j.talanta.2014.02.003.

[24] Olennikov, D.N., N.I. Kashchenko, N.K. Chirikova, S.S. Kuz'Mina, Phenolic profile of Potentilla anserina L. (Rosaceae) herb of Siberian origin and development of a rapid method for simultaneous determination of major phenolics in P. anserina pharmaceutical products by microcolumn RP-HPLC-UV, *Molecules.* 20 (2015) 224–248. doi: 10.3390/molecules20010224.

[25] Hornshøj, B.H., S. Kobbelgaard, W.R. Blakemore, H. Stapelfeldt, H.J. Bixler, M. Klinger, Quantification of free formaldehyde in carrageenan and processed Eucheuma seaweed using high-performance liquid chromatography, Food Addit. Contam. - Part A *Chem. Anal. Control. Expo. Risk Assess.* 32 (2015) 152–160. doi: 10.1080/19440049.2014.992049.

[26] Song, J., F. Chen, J. Liu, Y. Zou, Y. Luo, X. Yi, J. Meng, X. Chen, Combinative method using multi-components quantitation and HPLC fingerprint for comprehensive evaluation of Gentiana crassicaulis, Pharmacogn. *Mag.* 13 (2017) 180–187. doi: 10.4103/0973-1296. 197639.

[27] Meda, N.R., D. Fraisse, C. Gnoula, M. Vivier, C. Felgines, F. Senejoux, Characterization of antioxidants from Detarium microcarpum Guill. et Perr. leaves using HPLC-DAD coupled with pre-column DPPH assay, *Eur. Food Res. Technol.* 243 (2017) 1659–1666. doi: 10.1007/s00217-017-2873-7.

[28] Aysun Kepekçi, R., S. Polat, G. Çoşkun, A. Çelik, A.S. Bozkurt, Ö. Yumrutaş, M. Pehlivan, Preliminary Characterization of Phenolic Acid Composition and Hepatoprotective Effect of Stachys pumila, *J. Food Biochem.* 41 (2017). doi: 10.1111/jfbc.12286.

[29] Al-Rimawi, F., F. Alakhras, W.A. Al-Zereini, H.K. Aldal, S. Abubu-Lafi, G.M. Al-Mazaideh, H.J.A. Salman, HPLC analysis of chemical composition of selected jordanian medicinal plants and their bioactive properties, *Orient. J. Chem.* 34 (2018) 2397–2403. doi: 10.13005/ojc/340522.

[30] Singleton, C., R. Brkljača, S. Urban, Absolute configuration determination of retroflexanone using the advanced mosher method and application of HPLC-NMR, *Mar. Drugs.* 16 (2018) 1–10. doi: 10.3390/md16060205.

[31] Luo, C., F. Yi, Y. Xia, Z. Huang, X. Zhou, X. Jin, Y. Tang, J. Yi, Comprehensive quality evaluation of the lateral root of Aconitum carmichaelii Debx. (Fuzi): Simultaneous determination of nine alkaloids and chemical fingerprinting coupled with chemometric analysis, *J. Sep. Sci.* 42 (2019) 980–990. doi: 10.1002/jssc.201800937.

[32] Zeng, Q., Y. Ruan, L. Sun, F. Du, L. Guo, Z. Cheng, G. Ruan, J. Li, Development of Graphene Oxide Functionalized Cotton Fiber Based Solid Phase Extraction Combined with Liquid Chromatography-Fluorescence Detection for Determination of Trace Auxins in Plant Samples, *Chromatographia.* 81 (2018) 861–869. doi: 10.1007/s10337-018-3518-0.

[33] Sharifiyan, F., S.A. Mirjalili, M. Fazilati, E. Poorazizi, S. Habibollahi, Variation of ursolic acid content in flowers of ten Iranian pomegranate (Punica granatum L.) cultivars, *BMC Chem.* 13 (2019) 1–8. doi: 10.1186/s13065-019-0598-3.

[34] Sroka, Z., G. Zgórka, B. Zbikowska, A. Sowa, R. Franiczek, K. Wychowaniec, B. Krzyanowska, High antimicrobial efficacy, antioxidant activity, and a novel approach to phytochemical analysis of bioactive polyphenols in extracts from leaves of pyrus communis and pyrus pyrifolia collected during one vegetative season, *Microb. Drug Resist.* 25 (2019) 582–593. doi: 10.1089/mdr.2018.0149.

[35] Najafian, S., M. Moradi, M. Sepehrimanesh, Polyphenolic contents and antioxidant activities of two medicinal plant species, Mentha piperita and Stevia rebaudiana, cultivated in Iran, *Comp. Clin. Path.* 25 (2016) 743–747. doi: 10.1007/s00580-016-2258-5.

[36] González-Trujano, M.E., F. Pellicer, P. Mena, D.A. Moreno, C. García-Viguera, Antinociceptive and anti-inflammatory activities of a pomegranate (Punica granatum L.) extract rich in ellagitannins, *Int. J.*

Food Sci. Nutr. 66 (2015) 395–399. doi: 10.3109/09637486.2015. 1024208.

[37] Lin, Y.K., Y.L. Ho, Y. Zhao, Y.S. Chang, Quality assessment of Fritillariae Thunbergii Bulbus sold in Taiwan markets using a validated HPLC-UV method combined with hierarchical clustering analysis, *J. Food Drug Anal.* 23 (2015) 130–135. doi: 10.1016/j.jfda. 2014.06.004.

[38] Reza, H.M., Z.T. Gias, P. Islam, S. Sabnam, P. Jain, M.H. Hossain, M.A. Alam, HPLC-DAD system-based phenolic content analysis and in vitro antioxidant activities of rice bran obtained from aush dhan (Oryza sativa) of Bangladesh, *J. Food Biochem.* 39 (2015) 462–470. doi: 10.1111/jfbc.12154.

[39] Megeressa, M., D. Bisrat, A. Mazumder, K. Asres, Structural elucidation of some antimicrobial constituents from the leaf latex of Aloe trigonantha L.C. Leach, BMC Complement. *Altern. Med.* 15 (2015) 1–8. doi: 10.1186/s12906-015-0803-4.

[40] Zhou, X., L. Tang, H. Wu, G. Zhou, T. Wang, Z. Kou, S. Li, Z. Wang, Chemometric analyses for the characterization of raw and processed seeds of Descurainia sophia (L.) based on HPLC fingerprints, *J. Pharm. Biomed. Anal.* 111 (2015) 1–6. doi: 10.1016/j.jpba.2015. 03.010.

[41] Alvarez-Zapata, R., A. Sánchez-Medina, M. Chan-Bacab, K. García-Sosa, F. Escalante-Erosa, R.V. García-Rodríguez, L.M. Peña-Rodríguez, Chemometrics-enhanced high performance liquid chromatography-ultraviolet detection of bioactive metabolites from phytochemically unknown plants, *J. Chromatogr. A.* 1422 (2015) 213–221. doi: 10.1016/j.chroma.2015.10.026.

[42] Limtrakul, P., S. Yodkeeree, P. Thippraphan, W. Punfa, J. Srisomboon, Anti-aging and tyrosinase inhibition effects of Cassia fistula flower butanolic extract, BMC Complement. *Altern. Med.* 16 (2016) 1–10. doi: 10.1186/s12906-016-1484-3.

[43] Loizzo, M.R., V. Sicari, M.C. Tenuta, M.R. Leporini, T. Falco, T.M. Pellicanò, F. Menichini, R. Tundis, Phytochemicals content, antioxidant and hypoglycaemic activities of commercial nutmeg mace

(Myristica fragrans L.) and pimento (Pimenta dioica (L.) Merr.), *Int. J. Food Sci. Technol.* 51 (2016) 2057–2063. doi: 10.1111/ijfs.13178.

[44] Chan, C.L., R.Y. Gan, H. Corke, The phenolic composition and antioxidant capacity of soluble and bound extracts in selected dietary spices and medicinal herbs, *Int. J. Food Sci. Technol.* 51 (2016) 565–573. doi: 10.1111/ijfs.13024.

[45] Liang, X., C. Zhao, W. Su, A Fast and Reliable UPLC-PAD Fingerprint Analysis of Chimonanthus salicifolius Combined with Chemometrics Methods, *J. Chromatogr. Sci.* 54 (2016) 1213–1219. doi: 10.1093/chromsci/bmw053.

[46] Kendir, G., E. Dinç, A.K. Güvenç, Ultra-performance liquid chromatography for the simultaneous quantification of rutin and chlorogenic acid in leaves of Ribes L. Species by conventional and chemometric calibration approaches, *J. Chromatogr. Sci.* 53 (2015) 1577–1587. doi: 10.1093/chromsci/bmv060.

[47] Guillen Quispe, Y.N., S.H. Hwang, Z. Wang, S.S. Lim, Screening of Peruvian medicinal plants for tyrosinase inhibitory properties: Identification of tyrosinase inhibitors in Hypericum laricifolium juss, *Molecules*. 22 (2017). doi: 10.3390/molecules22030402.

[48] Li, M.N., X. Dong, W. Gao, X.G. Liu, R. Wang, P. Li, H. Yang, Global identification and quantitative analysis of chemical constituents in traditional Chinese medicinal formula Qi-Fu-Yin by ultra-high performance liquid chromatography coupled with mass spectrometry, *J. Pharm. Biomed. Anal.* 114 (2015) 376–389. doi: 10.1016/j.jpba.2015.05.030.

[49] Dong, Q., L.L. Qiu, C.E. Zhang, L.H. Chen, W.W. Feng, L.N. Ma, D. Yan, M. Niu, J. bo Wang, X. he Xiao, Identification of compounds in an anti-fibrosis Chinese medicine (Fufang Biejia Ruangan Pill) and its absorbed components in rat biofluids and liver by UPLC-MS, *J. Chromatogr. B Anal. Technol. Biomed. Life Sci.* 1026 (2016) 145–151. doi: 10.1016/j.jchromb.2015.12.024.

[50] de Hoffmann, E., V. Stroobant, Mass spectrometry: Principles and applications, 2007. https://linkinghub.elsevier.com/retrieve/pii/S0167924497800169.

[51] Ekman, R., J. Silberring, A. Westman-Brinkmalm, A. Kraj, MASS SPECTROMETRY Instrumentation, *Interpretation, and Applications*, 2009.

[52] Malisch, C.S., A. Lüscher, N. Baert, M.T. Engström, B. Studer, C. Fryganas, D. Suter, I. Mueller-Harvey, J.P. Salminen, Large Variability of Proanthocyanidin Content and Composition in Sainfoin (Onobrychis viciifolia), *J. Agric. Food Chem.* 63 (2015) 10234–10242. doi: 10.1021/acs.jafc.5b04946.

[53] Miura, M., M. Sakai, M. Nogami, M. Sato, T. Yatsushiro, A rapid LC–MS/MS method for lutein quantification in spinach (Spinacia oleracea), *Microchem. J.* 153 (2020) 104470. doi: 10.1016/j.microc.2019.104470.

[54] *Waters webpage*, Available from: https://www.waters.com/nextgen/xg/en/shop/columns/186002885-acquity-uplc-beh-phenyl-column-130a-17--m-21-mm-x-100-mm-1-pk.html, Accessed on 5th May 2020., (n.d.).

[55] Soufi, S., G. D'Urso, C. Pizza, S. Rezgui, T. Bettaieb, P. Montoro, Steviol glycosides targeted analysis in leaves of Stevia rebaudiana (Bertoni) from plants cultivated under chilling stress conditions, *Food Chem.* 190 (2016) 572–580. doi: 10.1016/j.foodchem.2015.05.116.

[56] Tessema, E.N., T. Gebre-Mariam, C.E.H. Schmelzer, R.H.H. Neubert, Isolation and structural characterization of glucosylceramides from Ethiopian plants by LC/APCI-MS/MS, *J. Pharm. Biomed. Anal.* 141 (2017) 241–249. doi: 10.1016/j.jpba.2017.04.036.

[57] *YMC webpage*, Available from: http://www.ymc.co.jp/en/columns/ymc_pack_ods_aq/, Accessed on 5th May 2020., (n.d.).

[58] Zhang, Y., L. Guo, L. Duan, X. Dong, P. Zhou, E.H. Liu, P. Li, Simultaneous determination of 16 phenolic constituents in Spatholobi Caulis by high performance liquid chromatography/electrospray ionization triple quadrupole mass spectrometry, *J. Pharm. Biomed. Anal.* 102 (2015) 110–118. doi: 10.1016/j.jpba.2014.09.006.

[59] Tang, S., S. Liu, Z. Liu, F. Song, S. Liu, Analysis and identification of the chemical constituents of Ding-Zhi-Xiao-Wan prescription by

HPLC-IT-MSn and HPLC-Q-TOF-MS, *Chinese J. Chem.* 33 (2015) 451–462. doi: 10.1002/cjoc.201400789.

[60] Musharraf, S.G., M. Goher, B. Zareena, Quantification of steroidal alkaloids in Buxus papillosa using electrospray ionization liquid chromatography-triple quadrupole mass spectrometry, *Steroids.* 100 (2015) 5–10. doi: 10.1016/j.steroids.2015.03.018.

[61] Zhao, W., X. Huang, X. Li, F. Zhang, S. Chen, M. Ye, M. Huang, W. Xu, S. Wu, Qualitative and quantitative analysis of major triterpenoids in alismatis rhizoma by high performance liquid chromatography/ diode-array detector/quadrupole- time-of-flight mass spectrometry and ultra-performance liquid chromatography/triple quadrupole mas, *Molecules.* 20 (2015) 13958–13981. doi: 10.3390/molecules200813958.

[62] Lin, Y., W. Xu, M. Huang, W. Xu, H. Li, M. Ye, X. Zhang, K. Chu, Qualitative and quantitative analysis of phenolic acids, flavonoids and iridoid glycosides in Yinhua Kanggan tablet by UPLC-QqQ-MS/MS, *Molecules.* 20 (2015) 12209–12228. doi: 10.3390/molecules200712209.

[63] Chen, D., S. Lin, W. Xu, M. Huang, J. Chu, F. Xiao, J. Lin, J. Peng, Qualitative and quantitative analysis of the major constituents in Shexiang Tongxin dropping pill by HPLC-Q-TOF-MS/MS and UPLC-QqQ-MS/MS, *Molecules.* 20 (2015) 18597–18619. doi: 10.3390/molecules201018597.

[64] Lin, Y., W. Xu, W. Xu, M. Huang, Y. Zhang, H. Li, K. Chu, L. Chen, Simultaneous determination of 41 components in Gualou Guizhi granules by UPLC coupled with triple quadrupole mass spectrometry, *Anal. Methods.* 7 (2015) 8285–8296. doi: 10.1039/c5ay01336d.

[65] Ali, A., S. Maher, S.A. Khan, M.I. Chaudhary, S.G. Musharraf, Sensitive quantification of six steroidal lactones in Withania coagulans extract by UHPLC electrospray tandem mass spectrometry, *Steroids.* 104 (2015) 176–181. doi: 10.1016/j.steroids.2015.09.011.

[66] Bell, L., M.J. Oruna-Concha, C. Wagstaff, Identification and quantification of glucosinolate and flavonol compounds in rocket salad (Eruca sativa, Eruca vesicaria and Diplotaxis tenuifolia) by LC-

MS: Highlighting the potential for improving nutritional value of rocket crops, *Food Chem.* 172 (2015) 852–861. doi: 10.1016/j.foodchem.2014.09.116.

[67] Wang, C., Y. Xie, Z. Xiang, H. Zhou, L. Liu, Simultaneous determination of thirteen major active compounds in Guanjiekang preparation by UHPLC-QQQ-MS/MS, *J. Pharm. Biomed. Anal.* 118 (2016) 315–321. doi: 10.1016/j.jpba.2015.10.043.

[68] Guo, L., S.L. Zeng, Y. Zhang, P. Li, E.H. Liu, Comparative analysis of steroidal saponins in four Dioscoreae herbs by high performance liquid chromatography coupled with mass spectrometry, *J. Pharm. Biomed. Anal.* 117 (2016) 91–98. doi: 10.1016/j.jpba.2015.08.038.

[69] Sun, Q., H. Cao, Y. Zhou, X. Wang, H. Jiang, L. Gong, Y. Yang, R. Rong, Qualitative and quantitative analysis of the chemical constituents in Mahuang-Fuzi-Xixin decoction based on high performance liquid chromatography combined with time-of-flight mass spectrometry and triple quadrupole mass spectrometers, *Biomed. Chromatogr.* 30 (2016) 1820–1834. doi: 10.1002/bmc.3758.

[70] Chen, J., Q. Chen, F. Yu, H. Huang, P. Li, J. Zhu, X. He, Comprehensive characterization and quantification of phillyrin in the fruits of Forsythia suspensa and its medicinal preparations by liquid chromatography-ion trap mass spectrometry, *Acta Chromatogr.* 28 (2016) 145–157. doi: 10.1556/AChrom.28.2016.1.11.

[71] Wang, J., L. Peng, M. Shi, C. Li, Y. Zhang, W. Kang, Spectrum Effect Relationship and Component Knock-Out in Angelica Dahurica Radix by High Performance Liquid Chromatography-Q Exactive Hybrid Quadrupole-Orbitrap Mass Spectrometer, *Molecules.* 22 (2017). doi: 10.3390/molecules22071231.

[72] Da Liu, G., Y.W. Zhao, Y.J. Li, X.J. Wang, H.H. Si, W.Z. Huang, Z.Z. Wang, S.P. Ma, W. Xiao, Qualitative and quantitative analysis of major constituents from Dazhu Hongjingtian capsule by UPLC/Q-TOF-MS/MS combined with UPLC/QQQ-MS/MS, *Biomed. Chromatogr.* 31 (2017) 1–10. doi: 10.1002/bmc.3887.

[73] Jia, S., J. Song, R. Dai, Y. Deng, F. Lv, Structural deduction of pregnane glycosides from Dregea sinensis Hemsl by high-

performance liquid chromatography and electrospray ionization-tandem mass spectrometry, *Int. J. Mass Spectrom.* 415 (2017) 38–43. doi: 10.1016/j.ijms.2017.01.018.

[74] Li, S., S. Liu, Z. Pi, F. Song, Y. Jin, Z. Liu, Chemical profiling of Fufang-Xialian-Capsule by UHPLC-Q-TOF-MS and its antioxidant activity evaluated by in vitro method, *J. Pharm. Biomed. Anal.* 138 (2017) 289–301. doi: 10.1016/j.jpba.2017.01.060.

[75] Pu, C.H., S.K. Lin, W.C. Chuang, T.H. Shyu, Modified QuEChERS method for 24 plant growth regulators in grapes using LC-MS/MS, *J. Food Drug Anal.* 26 (2018) 637–648. doi: 10.1016/j.jfda.2017.08.001.

[76] Yilmaz, M.A., A. Ertas, I. Yener, M. Akdeniz, O. Cakir, M. Altun, I. Demirtas, M. Boga, H. Temel, A comprehensive LC–MS/MS method validation for the quantitative investigation of 37 fingerprint phytochemicals in Achillea species: A detailed examination of A. coarctata and A. monocephala, *J. Pharm. Biomed. Anal.* 154 (2018) 413–424. doi: 10.1016/j.jpba.2018.02.059.

[77] Sixto, A., A. Pérez-Parada, S. Niell, H. Heinzen, GC–MS and LC–MS/MS workflows for the identification and quantitation of pyrrolizidine alkaloids in plant extracts, a case study: Echium plantagineum, *Brazilian J. Pharmacogn.* 29 (2019) 500–503. doi: 10.1016/j.bjp.2019.04.010.

[78] Huang, Y., A.S. Adeleye, L. Zhao, A.S. Minakova, T. Anumol, A.A. Keller, Antioxidant response of cucumber (Cucumis sativus) exposed to nano copper pesticide: Quantitative determination via LC-MS/MS, *Food Chem.* 270 (2019) 47–52. doi: 10.1016/j.foodchem.2018.07.069.

[79] Ersoy, E., E. Eroglu Ozkan, M. Boga, M.A. Yilmaz, A. Mat, Anti-aging potential and anti-tyrosinase activity of three Hypericum species with focus on phytochemical composition by LC–MS/MS, *Ind. Crops Prod.* 141 (2019). doi: 10.1016/j.indcrop.2019.111735.

[80] Mighri, H., A. Akrout, N. Bennour, H. Eljeni, T. Zammouri, M. Neffati, LC/MS method development for the determination of the phenolic compounds of Tunisian Ephedra alata hydro-methanolic extract and its fractions and evaluation of their antioxidant activities,

South African J. Bot. 124 (2019) 102–110. doi: 10.1016/j.sajb.2019.04.029.

[81] Sarikurkcu, C., M.S. Ozer, N. Tlili, LC–ESI–MS/MS characterization of phytochemical and enzyme inhibitory effects of different solvent extract of Symphytum anatolicum, *Ind. Crops Prod.* 140 (2019) 111666. doi: 10.1016/j.indcrop.2019.111666.

[82] Ozdal, T., F.D. Ceylan, N. Eroglu, M. Kaplan, E.O. Olgun, E. Capanoglu, Investigation of antioxidant capacity, bioaccessibility and LC-MS/MS phenolic profile of Turkish propolis, *Food Res. Int.* 122 (2019) 528–536. doi: 10.1016/j.foodres.2019.05.028.

[83] Cherfia, R., A. Zaiter, S. Akkal, P. Chaimbault, A.B. Abdelwahab, G. Kirsch, N. Kacem Chaouche, New approach in the characterization of bioactive compounds isolated from Calycotome spinosa (L.) Link leaves by the use of negative electrospray ionization LITMSn, LC-ESI-MS/MS, as well as NMR analysis, *Bioorg. Chem.* 96 (2020) 103535. doi: 10.1016/j.bioorg.2019.103535.

[84] Seo, C.S., H.K. Shin, Quality assessment of traditional herbal formula, Hyeonggaeyeongyo-tang through simultaneous determi-nation of twenty marker components by HPLC–PDA and LC–MS/MS, *Saudi Pharm. J.* (2020). doi:10.1016/j.jsps.2020.02.003.

[85] Khan, M.N., F. Ul Haq, S. Rahman, A. Ali, S.G. Musharraf, Metabolite distribution and correlation studies of Ziziphus jujuba and Ziziphus nummularia using LC-ESI-MS/MS, *J. Pharm. Biomed. Anal.* 178 (2020) 112918. doi: 10.1016/j.jpba.2019.112918.

[86] Gülçin, İ., A.C. Gören, P. Taslimi, S.H. Alwasel, O. Kılıc, E. Bursal, Anticholinergic, antidiabetic and antioxidant activities of Anatolian pennyroyal (Mentha pulegium)-analysis of its polyphenol contents by LC-MS/MS, *Biocatal. Agric. Biotechnol.* 23 (2020). doi: 10.1016/j.bcab.2019.101441.

[87] Yilmaz, M.A., Industrial Crops & Products Simultaneous quantitative screening of 53 phytochemicals in 33 species of medicinal and aromatic plants : A detailed, robust and comprehensive LC – MS/MS method validation, *Ind. Crop. Prod.* 149 (2020) 112347. doi: 10.1016/j.indcrop.2020.112347.

[88] Yin, J., W. Ren, B. Wei, H. Huang, M. Li, X. Wu, A. Wang, Z. Xiao, J. Shen, Y. Zhao, F. Du, H. Ji, P.J. Kaboli, Y. Ma, Z. Zhang, C.H. Cho, S. Wang, X. Wu, Y. Wang, Characterization of chemical composition and prebiotic effect of a dietary medicinal plant Penthorum chinense Pursh, *Food Chem.* 319 (2020). doi: 10.1016/j.foodchem.2020.126568.

[89] Liang, X., N.J. Nielsen, J.H. Christensen, Selective pressurized liquid extraction of plant secondary metabolites: Convallaria majalis L. as a case, *Anal. Chim. Acta X.* 4 (2020). doi: 10.1016/j.acax.2020.100040.

[90] Salem, M.A., M.M. Farid, M. El-Shabrawy, R. Mohammed, S.R. Hussein, M.M. Marzouk, Spectrometric analysis, chemical constituents and cytotoxic evaluation of Astragalus sieberi DC. (Fabaceae), *Sci. African.* 7 (2020) e00221. doi: 10.1016/j.sciaf.2019.e00221.

[91] Enneb, S., S. Drine, M. Bagues, T. Triki, F. Boussora, F. Guasmi, K. Nagaz, A. Ferchichi, Phytochemical profiles and nutritional composition of squash (Cucurbita moschata D.) from Tunisia, *South African J. Bot.* 130 (2020) 165–171. doi: 10.1016/j.sajb.2019.12.011.

[92] Wang, C.J., F. He, Y.F. Huang, H.L. Ma, Y.P. Wang, C.S. Cheng, J. Le Cheng, C.C. Lao, D.A. Chen, Z.F. Zhang, Z. Sang, P. Luo, S.Y. Xiao, Y. Xie, H. Zhou, Discovery of chemical markers for identifying species, growth mode and production area of Astragali Radix by using ultra-high-performance liquid chromatography coupled to triple quadrupole mass spectrometry, *Phytomedicine.* 67 (2020). doi: 10.1016/j.phymed.2019.153155.

[93] Liu, J., M. Wang, L. Chen, Y. Li, Y. Chen, Z. Wei, Z. Jia, W. Xu, H. Xiao, Profiling the constituents of Dachuanxiong decoction by liquid chromatography with high-resolution tandem mass spectrometry using target and nontarget data mining, *J. Sep. Sci.* 42 (2019) 2202–2213. doi: 10.1002/jssc.201900064.

[94] Wolfender, J.L., G. Marti, A. Thomas, S. Bertrand, Current approaches and challenges for the metabolite profiling of complex natural extracts, *J. Chromatogr. A.* 1382 (2015) 136–164. doi: 10.1016/j.chroma.2014.10.091.

[95] Purves, R.W., Enhancement of biological mass spectrometry by using separations based on changes in ion mobility (FAIMS and DMS), *Anal. Bioanal. Chem.* 405 (2013) 35–42. doi: 10.1007/s00216-012-6496-3.

[96] Ganzera, M., S. Sturm, Recent advances on HPLC/MS in medicinal plant analysis—An update covering 2011–2016, *J. Pharm. Biomed. Anal.* 147 (2018) 211–233. doi: 10.1016/j.jpba.2017.07.038.

[97] Capriotti, A.L., C. Cavaliere, G. La Barbera, C.M. Montone, S. Piovesana, R. Zenezini Chiozzi, A. Laganà, Chromatographic column evaluation for the untargeted profiling of glucosinolates in cauliflower by means of ultra-high performance liquid chromato-graphy coupled to high resolution mass spectrometry, *Talanta.* 179 (2018) 792–802. doi: 10.1016/j.talanta.2017.12.019.

[98] Sinitsyn, M.Y., A.V. Aksenov, M.V. Taranchenko, I.A. Rodin, A.N. Stavrianidi, A.M. Antokhin, O.A. Shpigun, Structural Characterization of Triterpene Saponins from Manchurian Aralia by High Resolution Liquid Chromatography–Mass Spectrometry, *J. Anal. Chem.* 74 (2019) 1113–1121. doi: 10.1134/S1061934819110108.

[99] Li, Y., J. Liu, R. Su, Q. Li, Y. Chen, J. Yang, S. Zhao, Z. Jia, H. Xiao, Pseudotargeted screening and determination of constituents in Qishen granule based on compound biosynthetic correlation using UHPLC coupled with high-resolution MS, *J. Sep. Sci.* (2020) 1–11. doi: 10.1002/jssc.201900980.

[100] Gao, X., N. Wang, J. Jia, P. Wang, A. Zhang, X. Qin, Chemical profliling of Dingkun Dan by ultra High performance liquid chromatography Q exactive orbitrap high resolution mass spectrometry, *J. Pharm. Biomed. Anal.* 177 (2020) 112732. doi: 10.1016/j.jpba.2019.06.029.

[101] Lima, R. de C.L., S.M. Gramsbergen, J. Van Staden, A.K. Jäger, K.T. Kongstad, D. Staerk, Advancing HPLC-PDA-HRMS-SPE-NMR Analysis of Coumarins in Coleonema album by Use of Orthogonal Reversed-Phase C18 and Pentafluorophenyl Separations, *J. Nat. Prod.* 80 (2017) 1020–1027. doi: 10.1021/acs.jnatprod.6b01020.

[102] Núñez, O., A. Checa, H. Gallart-Ayala, Fluorinated Stationary Phases on Liquid Chromatography: Preparation, Properties and Applications, in F. Ramos (ed.) *Liquid Chromatography: Principles, Technology and Applications.*, Nova Science Publishers, Inc., New York, USA, 2013.

[103] Ballesteros-Vivas, D., G. Álvarez-Rivera, A. del Pilar Sánchez-Camargo, E. Ibáñez, F. Parada-Alfonso, A. Cifuentes, A multi-analytical platform based on pressurized-liquid extraction, in vitro assays and liquid chromatography/gas chromatography coupled to high resolution mass spectrometry for food by-products valorisation. Part 1: Withanolides-rich fractions from golde, *J. Chromatogr. A.* 1584 (2019) 155–164. doi: 10.1016/j.chroma.2018.11.055.

[104] Tao, S., Y. Huang, Z. Chen, Y. Chen, Y. Wang, Rapid identification of anti-inflammatory compounds from Tongmai Yangxin Pills by liquid chromatography with high-resolution mass spectrometry and chemometric analysis, *J. Sep. Sci.* 38 (2015) 1881–1893. doi: 10.1002/jssc.201401481.

[105] Tahtah, Y., K.T. Kongstad, S.G. Wubshet, N.T. Nyberg, L.H. Jønsson, A.K. Jäger, S. Qinglei, D. Staerk, Triple aldose reductase/α-glucosidase/radical scavenging high-resolution profiling combined with high-performance liquid chromatography-high-resolution mass spectrometry-solid-phase extraction-nuclear magnetic resonance spectroscopy foridentification of a, *J. Chromatogr. A.* 1408 (2015) 125–132. doi: 10.1016/j.chroma.2015.07.010.

[106] Citti, C., G. Ciccarella, D. Braghiroli, C. Parenti, M.A. Vandelli, G. Cannazza, Medicinal cannabis: Principal cannabinoids concentration and their stability evaluated by a high performance liquid chromatography coupled to diode array and quadrupole time of flight mass spectrometry method, *J. Pharm. Biomed. Anal.* 128 (2016) 201–209. doi: 10.1016/j.jpba.2016.05.033.

[107] Iqbal, K., J. Iqbal, D. Staerk, K.T. Kongstad, Characterization of antileishmanial compounds from Lawsonia inermis L. Leaves using semi-high resolution antileishmanial profiling combined with HPLC-

HRMS-SPE-NMR, *Front. Pharmacol.* 8 (2017) 1–7. doi: 10.3389/fphar.2017.00337.

[108] He, M., H. Wu, J. Nie, P. Yan, T.B. Yang, Z.Y. Yang, R. Pei, Accurate recognition and feature qualify for flavonoid extracts from Liang-wai Gan Cao by liquid chromatography-high resolution-mass spectrometry and computational MS/MS fragmentation, *J. Pharm. Biomed. Anal.* 146 (2017) 37–47. doi: 10.1016/j.jpba.2017.07.065.

[109] Shin, H., Y. Park, Y.H. Jeon, X.T. Yan, K.Y. Lee, Identification of Polygonum orientale constituents using high-performance liquid chromatography high-resolution tandem mass spectrometry, *Biosci. Biotechnol. Biochem.* 82 (2018) 15–21. doi: 10.1080/09168451.2017.1415124.

[110] Ping Huang, W., T. Tan, Z. feng Li, H. OuYang, X. Xu, B. Zhou, Y. lin Feng, Structural characterization and discrimination of Chimonanthus nitens Oliv. leaf from different geographical origins based on multiple chromatographic analysis combined with chemometric methods, *J. Pharm. Biomed. Anal.* 154 (2018) 236–244. doi: 10.1016/j.jpba.2018.02.020.

[111] Brakni, R., M. Ali Ahmed, P. Burger, A. Schwing, G. Michel, C. Pomares, L. Hasseine, L. Boyer, X. Fernandez, A. Landreau, T. Michel, UHPLC-HRMS/MS Based Profiling of Algerian Lichens and Their Antimicrobial Activities, *Chem. Biodivers.* 15 (2018). doi: 10.1002/cbdv.201800031.

[112] Rivera-Mondragón, A., S. Bijttebier, E. Tuenter, D. Custers, O.O. Ortíz, L. Pieters, C. Caballero-George, S. Apers, K. Foubert, Phytochemical characterization and comparative studies of four Cecropia species collected in Panama using multivariate data analysis, *Sci. Rep.* 9 (2019) 1–14. doi: 10.1038/s41598-018-38334-4.

[113] Geoffroy, T.R., T. Stevanovic, Y. Fortin, P.E. Poubelle, N.R. Meda, Metabolite Profiling of Two Maple-Derived Products Using Dereplication Based on High-Performance Liquid Chromatography-Diode Array Detector-Electrospray Ionization-Time-of-Flight-Mass Spectrometry: Sugar Maple Bark and Bud Hot-Water Extracts, *J.*

Agric. Food Chem. 67 (2019) 8819–8838. doi: 10.1021/acs.jafc. 9b02664.

[114] He, M., G. Peng, F. Xie, L. Hong, Q. Cao, Liquid Chromatography–High-Resolution Mass Spectrometry with ROI Strategy for Nontargeted Analysis of the In Vivo/In Vitro Ingredients Coming from Ligusticum chuanxiong hort, *Chromatographia.* 82 (2019) 1069–1077. doi: 10.1007/s10337-019-03740-x.

[115] Yu, C., Y. Xu, M. Wang, Z. Xie, X. Gao, Application of characteristic fragment filtering with ultra high performance liquid chromatography coupled with high-resolution mass spectrometry for comprehensive identification of components in Schisandrae chinensis Fructus, *J. Sep. Sci.* 42 (2019) 1323–1331. doi: 10.1002/jssc.201801203.

[116] Zhao, N., M. Cheng, W. Lv, Y. Wu, D. Liu, X. Zhang, Peptides as Potential Biomarkers for Authentication of Mountain-Cultivated Ginseng and Cultivated Ginseng of Different Ages Using UPLC-HRMS, *J. Agric. Food Chem.* (2020). doi: 10.1021/acs.jafc.9b05568.

[117] Delgado-Povedano, M.M., C. Sánchez-Carnerero Callado, F. Priego-Capote, C. Ferreiro-Vera, Untargeted characterization of extracts from Cannabis sativa L. cultivars by gas and liquid chromatography coupled to mass spectrometry in high resolution mode, *Talanta.* 208 (2019) 120384. doi: 10.1016/j.talanta.2019.120384.

[118] Cao, G., X. Chen, X. Wu, Q. Li, H. Zhang, Rapid identification and comparative analysis of chemical constituents in herbal medicine Fufang decoction by ultra-high-pressure liquid chromatography coupled with a hybrid linear ion trap-high-resolution mass spectrometry, *Biomed. Chromatogr.* 29 (2015) 698–708. doi: 10.1002/bmc.3333.

[119] Simirgiotis, M.J., C. Quispe, J. Bórquez, A. Mocan, B. Sepúlveda, High resolution metabolite fingerprinting of the resin of Baccharis tola Phil. from the Atacama Desert and its antioxidant capacities, *Ind. Crops Prod.* 94 (2016) 368–375. doi: 10.1016/j.indcrop.2016.08.037.

[120] Xu, T., L. Zuo, Z. Sun, P. Wang, L. Zhou, X. Lv, Q. Jia, X. Liu, X. Jiang, Z. Zhu, J. Kang, X. Zhang, Chemical profiling and quantification of ShenKang injection, a systematic quality control

strategy using ultra high performance liquid chromatography with Q Exactive hybrid quadrupole orbitrap high-resolution accurate mass spectrometry, *J. Sep. Sci.* 40 (2017) 4872–4879. doi: 10.1002/jssc. 201700928.

[121] Leitão, S.G., G.G. Leitão, D.K.T. Vicco, J.P.B. Pereira, G. de Morais Simão, D.R. Oliveira, R. Celano, L. Campone, A.L. Piccinelli, L. Rastrelli, Counter-current chromatography with off-line detection by ultra high performance liquid chromatography/high resolution mass spectrometry in the study of the phenolic profile of Lippia origanoides, *J. Chromatogr. A.* 1520 (2017) 83–90. doi: 10.1016/j.chroma.2017.09.004.

[122] Lee, S.T., D.R. Gardner, D. Cook, Identification of Indole Diterpenes in Ipomoea asarifolia and Ipomoea muelleri, Plants Tremorgenic to Livestock, *J. Agric. Food Chem.* 65 (2017) 5266–5277. doi: 10.1021/acs.jafc.7b01834.

[123] Yu, F., H. Qian, J. Zhang, J. Sun, Z. Ma, Simultaneous quantification of eight organic acid components in Artemisia capillaris Thunb (Yinchen) extract using high-performance liquid chromatography coupled with diode array detection and high-resolution mass spectrometry, *J. Food Drug Anal.* 26 (2018) 788–795. doi: 10.1016/j.jfda.2017.04.003.

[124] Sun, Z., Z. Li, L. Zuo, Z. Wang, L. Zhou, Y. Shi, J. Kang, Z. Zhu, X. Zhang, Qualitative and quantitative determination of YiXinShu Tablet using ultra high performance liquid chromatography with Q Exactive hybrid quadrupole orbitrap high-resolution accurate mass spectrometry, *J. Sep. Sci.* 40 (2017) 4453–4466. doi: 10.1002/jssc.201700619.

[125] Jiang, Y., R. Liu, M. Liu, L. Yi, S. Liu, An integrated strategy to rapidly characterize non-targeted benzylisoquinoline alkaloids from Plumula nelumbinis ethanol extract using UHPLC/Q-orbitrap HRMS, *Int. J. Mass Spectrom.* 432 (2018) 26–35. doi: 10.1016/j.ijms.2018.06.002.

[126] le Wang, P., Z. Sun, X. jing Lv, T. ye Xu, Q. quan Jia, X. Liu, X. fang Zhang, Z. feng Zhu, X. jian Zhang, A homologues prediction strategy

for comprehensive screening and characterization of C21 steroids from Xiao-ai-ping injection by using ultra high performance liquid chromatography coupled with high resolution hybrid quadrupole-orbitrap mass spectrometry, *J. Pharm. Biomed. Anal.* 148 (2018) 80–88. doi: 10.1016/j.jpba.2017.09.024.

[127] Zanatta, A.C., A. Mari, M. Masullo, I. Zeppone Carlos, W. Vilegas, S. Piacente, L. Campaner dos Santos, Chemical metabolome assay by high-resolution Orbitrap mass spectrometry and assessment of associated antitumoral activity of Actinocephalus divaricatus, *Rapid Commun. Mass Spectrom.* 32 (2018) 241–250. doi: 10.1002/rcm.8034.

[128] Lv, X.J., Z. Sun, P. Le Wang, J. Yang, T.Y. Xu, Q.Q. Jia, D.W. Li, F.Y. Su, Z.F. Zhu, J. Kang, X.J. Zhang, Chemical profiling and quantification of Dan-Deng-Tong-Nao-capsule using ultra high performance liquid chromatography coupled with high resolution hybrid quadruple-orbitrap mass spectrometry, *J. Pharm. Biomed. Anal.* 148 (2018) 189–204. doi: 10.1016/j.jpba.2017.09.034.

[129] Monga, G.K., A. Ghosal, D. Ramanathan, To develop the method for UHPLC-HRMS to determine the antibacterial potential of a central American medicinal plant, *Separations.* 6 (2019). doi: 10.3390/separations6030037.

[130] Yang, B., Y. Ying, J. Zou, S. Ge, L. Zuo, Comprehensive characterization and quantification of multiple components in Dan-Huang-Qu-Yu capsule using a multivariate data processing approach based on microwave-assisted extraction with UHPLC and Q Exactive quadrupole-orbitrap high-resolution mass spe, *J. Sep. Sci.* 42 (2019) 2069–2079. doi: 10.1002/jssc.201801246.

[131] Shakeri, A., G. D'Urso, S.F. Taghizadeh, S. Piacente, S. Norouzi, V. Soheili, J. Asili, D. Salarbashi, LC-ESI/LTQOrbitrap/MS/MS and GC–MS profiling of Stachys parviflora L. and evaluation of its biological activities, *J. Pharm. Biomed. Anal.* 168 (2019) 209–216. doi: 10.1016/j.jpba.2019.02.018.

[132] Chan, Ch-on., X. juan Xie, S. wai Wan, G. li Zhou, A.C. ying Yuen, D.K. wah Mok, S. bao Chen, Qualitative and quantitative analysis of

sesquiterpene lactones in Centipeda minima by UPLC–Orbitrap–MS & UPLC-QQQ-MS, *J. Pharm. Biomed. Anal.* 174 (2019) 360–366. doi: 10.1016/j.jpba.2019.05.067.

[133] Nijat, D., R. Abdulla, G. yu Liu, Y. qin Luo, H.A. Aisa, Identification and quantification of Meiguihua oral solution using liquid chromatography combined with hybrid quadrupole-orbitrap and triple quadrupole mass spectrometers, *J. Chromatogr. B Anal. Technol. Biomed. Life Sci.* 1139 (2020) 121992. doi: 10.1016/j.jchromb.2020.121992.

[134] D'Urso, G., P. Montoro, S. Piacente, Detection and comparison of phenolic compounds in different extracts of black currant leaves by liquid chromatography coupled with high-resolution ESI-LTQ-Orbitrap MS and high-sensitivity ESI-Qtrap MS, *J. Pharm. Biomed. Anal.* 179 (2020). doi: 10.1016/j.jpba.2019.112926.

[135] Guo, C., L. Gong, W. Wang, J. Leng, L. Zhou, S. Xing, Y. Zhao, R. Xian, X. Zhang, F. Shi, Rapid screening and identification of targeted or non-targeted antitussive adulterants in herbal medicines by Q-Orbitrap HRMS and screening database, *Int. J. Mass Spectrom.* 447 (2020) 116250. doi: 10.1016/j.ijms.2019.116250.

[136] Núñez, O., H. Gallart-Ayala, C.P.B. Martins, P. Lucci, New trends in fast liquid chromatography for food and environmental analysis, *J. Chromatogr. A.* 1228 (2012) 298–323. doi: 10.1016/j.chroma.2011.10.091.

[137] Núñez, O., H. Gallart-Ayala, C. Martins P B, P. Lucci, *Fast Liquid Chromatography– Mass Spectrometry Methods in Food and Environmental Analysis*, Imperial College Press, London, UK, 2015.

[138] Gosetti, F., E. Mazzucco, D. Zampieri, M.C. Gennaro, Signal suppression/enhancement in high-performance liquid chromatography tandem mass spectrometry, *J. Chromatogr. A.* 1217 (2010) 3929–3937. doi: 10.1016/j.chroma.2009.11.060.

[139] Kiontke, A., A. Oliveira-Birkmeier, A. Opitz, C. Birkemeyer, Electrospray ionization efficiency is dependent on different molecular descriptors with respect to solvent pH and instrumental configuration, *PLoS One.* 11 (2016) 1–16. doi: 10.1371/journal.pone.0167502.

[140] George, R., A. Haywood, S. Khan, M. Radovanovic, J. Simmonds, R. Norris, Enhancement and suppression of ionization in drug analysis using HPLC-MS/MS in support of therapeutic drug monitoring: A review of current knowledge of its minimization and assessment, *Ther. Drug Monit.* 40 (2018) 1–8. doi: 10.1097/FTD. 0000000000000471.

[141] Medeiros, R.M.T., R.C. Barbosa, F. Riet-Correa, E.F. Lima, I.M. Tabosa, S.S. De Barros, D.R. Gardner, R.J. Molyneux, Tremorgenic syndrome in goats caused by Ipomoea asarifolia in Northeastern Brazil, *Toxicon.* 41 (2003) 933–935. doi: 10.1016/S0041-0101(03) 00044-8.

[142] Zhang, H., J.M. Jiang, D. Zheng, M. Yuan, Z.Y. Wang, H.M. Zhang, C.W. Zheng, L.B. Xiao, H.X. Xu, A multidimensional analytical approach based on time-decoupled online comprehensive two-dimensional liquid chromatography coupled with ion mobility quadrupole time-of-flight mass spectrometry for the analysis of ginsenosides from white and red ginsengs, *J. Pharm. Biomed. Anal.* 163 (2019) 24–33. doi: 10.1016/j.jpba.2018.09.036.

[143] Venter, P., M. Muller, J. Vestner, M.A. Stander, A.G.J. Tredoux, H. Pasch, A. De Villiers, Comprehensive Three-Dimensional LC × LC × Ion Mobility Spectrometry Separation Combined with High-Resolution MS for the Analysis of Complex Samples, *Anal. Chem.* 90 (2018) 11643–11650. doi: 10.1021/acs.analchem.8b03234.

[144] Muller, M., A.G.J. Tredoux, A. de Villiers, Application of Kinetically Optimised Online HILIC × RP-LC Methods Hyphenated to High Resolution MS for the Analysis of Natural Phenolics, *Chromatographia.* 82 (2019) 181–196. doi: 10.1007/s10337-018-3662-6.

[145] liang Yao, C., W. zhi Yang, W. Si, Y. Shen, N. xia Zhang, H. li Chen, H. qin Pan, M. Yang, W. ying Wu, D. an Guo, An enhanced targeted identification strategy for the selective identification of flavonoid O-glycosides from Carthamus tinctorius by integrating offline two-dimensional liquid chromatography/linear ion-trap-Orbitrap mass

spectrometry, high-resolution diag, *J. Chromatogr. A.* 1491 (2017) 87–97. doi: 10.1016/j.chroma.2017.02.041.

[146] Schoenmakers, P., P. Marriott, J. Beens, Nomenclature and conventions incomprehensive multidimensional chromatography, *LC GC Eur.* 16 (2003) 335–339.

In: A Comprehensive Guide ...
Editors: Silje A. Dahl et al.

ISBN: 978-1-53618-418-1
© 2020 Nova Science Publishers, Inc.

Chapter 2

PHARMACOLOGICAL INTEREST OF CETRARIOID LICHENS IN THE PREVENTION OF OXIDATIVE STRESS-RELATED DISEASES

Isabel Ureña-Vacas, Elena González-Burgos[*]
and M. Pilar Gómez-Serranillos

Department of Pharmacology, Pharmacognosy and Botany,
Faculty of Pharmacy, University Complutense of Madrid, Spain

ABSTRACT

Lichens are symbiotic organisms composed by a mycobiont (fungus) and a photobiont (unicellular algae or cyanobacteria). Recent studies have revealed multispecific symbiosis with host-specific bacterial microbiomes. It is estimated that there are more than 28,000 lichen species worldwide among which Lecanorales order being the most abundant and including as the largest family Parmealiaceae, highlighting within it Cetrarioid clade.

Oxidative stress occurs when there is a disturbance between reactive oxygen species (ROS) and the antioxidant system (enzymatic and non-enzymatic), which may contribute to pathological chronic conditions

---

[*] Corresponding Author's Email: elenagon@ucm.es.

including cancer, cardiovascular diseases and neurodegenerative disorders. Exogenous antioxidants administration is considered the most promising strategy to cope oxidative stress based on its capacity to inhibit ROS action, chelate metal ions and increase enzyme activity and expression. At present, the interest in searching natural compounds as new antioxidants is increasing. Experimental studies have demonstrated that cetrarioid lichens and its major secondary metabolites are potential antioxidants with interest to prevent oxidative stress related diseases.

This chapter aims to provide a comprehensive overview of the pharmacological activity of cetrarioid lichens and its major secondary metabolites as antioxidants to prevent and treat oxidative stress-related diseases.

Keywords: lichen, cetrarioid clade, oxidative stress, antioxidants

INTRODUCTION

Lichens are symbiotic organisms composed by a mycobiont (fungus) and a photobiont (unicellular algae or cyanobacteria). The mycobiont allows algae live in terrestrial habitats providing water and simulating more humid and dismal environment. On the other hand, algae produce necessary nutrients for fungus through photosynthesis.

Recent studies have shown that a third component can interact with this dual symbiosis: specific bacterial microbiomes. They coexist with the mycobiont and live in association with the photobiont (multispecific symbiosis). The presence of these specific bacterial microbiomes influences in the composition of lichens and they are responsible for the synthesis of certain compounds and biological functions. They contain genes related to the development of biological function such as nutritional, hormonal regulation, defense against pathogens, tolerance to stress, detoxification and degradation of lichen thallus [1-4].

Morphologically, lichens are constituted by a thallus which can be homomer (homogeneous throughout its thallus) or heteromer (stratification in various layers). From the exterior to the interior, strata are classified into a) epicortex which is an acellular layer composed with polysaccharides secreted by hyphae; b) algal layer whose thickness varies according to the species and c) medulla, which occupies the majority of the thallus, and it has large storage capacity for water and metabolites as well as oxalate crystals in some lichen species [5-7].

It is estimated that there are more than 28,000 lichen species worldwide. Most lichenized fungi are ascomycetes that belong mainly to Lecanorales order that comprises 20 families [8]. Parmeliaceae family is the largest one of lichenized fungi (80 genera, 2,800 species), and within this, cetrarioid clade stands out phylogenetically by number. Recent phylogenetic studies based on DNA sequences have determined that cetrarioid clade is constituted by five subclades: *Cetraria, Nephromopsis, Dactylinia, Melaniela* and *Esslingeriana*.

CETRARIOID CLADE

Phylogeny, Morphology and Distribution

Cetraria clade is constituted by 5 genera (*Allocetraria* spp., *Cetraria* spp., *Cetrariella* spp., *Usnocetraria* spp., and *Vulpicida* spp.). The species of *Allocetraria* genus are characterized by foliaceous and fruticulous lichen thallus, yellowish-white medulla and, long and narrow conidia. These terrestrial lichen species are distributed in high altitudes of Asian countries (India, Nepal and China). Some of them such as *A. ambigua* are endemic of Himalayan vegetation. The species of *Cetraria* genus are also terrestrial lichens; they present fruticulosus thallus, mainly brown; white medulla and absent or marginal apothecia. *Cetraria* spp. are distributed in countries of Asian, Oceanic, European and American continents. On the other hand, species of *Cetrariella* genus have foliaceous (*C. commixta*) or fruticulosus (*C. delisei* and *C. fastigiata*) brown thallus and they live in diverse habitat. They can be found growing on rocks at Arctic-alpine level (*C. commixta*), on soils with dead leaves or humus (*C. delisei*) as well as at open humid sites in depressions of tundra and swamps (*C. fastigiata*) [8-11]. *Usnocetraria* species have yellowish-greenish foliaceous thallus and they live in spruce forests in the Northern Hemisphere. Finally, species of *Vulpicida* genus have foliaceous thallus and they live in mountainous lands on the bark of the trees [8, 12, 13].

Nephromopsis clade reunites 9 genera (*Ahtiana, Arctocetraria, Cetreliopsis, Flavocetraria, Kaernefeltia, Masonhalea, Nephromopsis, Tuckermanella* and *Tuckermannopsis*). Some species belonging to these genera are highlighted in this chapter. *Ahtiana pallidula* is endemic of North America; it presents yellowish-green foliaceous thallus and globose ascospores [14]. The specie *Arctocetraria nigricasens* is located in the Arctic tundra and it has yellowish-brown foliaceous thallus with rare cilia [15]. *Cetreliopsis rhytidocarpa*, which grows in East and Southeast Asia, has foliaceous thallus with pseudocephalodia on both surfaces [8]. Two species that compose the genus *Flavocetraria* are *F. cucullata* and *F. nivalis* and they have fruticolous yellow-greenish thallus and conidia with dumbbell form. These lichen species grow on dead leaves and mosses, mainly in the Northern Hemisphere [8]. The specie *Kaernefeltia merrillii* has blackish foliaceous thallus and white medulla and it is distributed in western North America and central Spain [8]. The genus *Masonhalea*, with two species presented in north of Beringia, is characterized by the presence of greenish-brown fructiculous (*M. inermis*) and greenish-leaved thallus (*M. richardsonii*). The genus *Nephromopsis* is constituted by a wide amount of species like *N. komarovii, N. laureri, N. leucostigma, N. nephromoides* and *N. pallecens*. *N. komarovii* presents foliaceous thallus attached to the substrate. *N. laureri* grows in Asia, Central and South America; it is characterized by a thallus with marginal soredia and absent isidia. The terrestrial species *N. leucostigma* has yellow laciniate thallus on its upper surface and brown on the lower one with cephalodia; it is distributed in Himalayan region. *N. nephromoides* has a slightly wrinkled thallus with cephalodia located on the lower surface and finally, *N. pallecens* is characterized by a small apothecium [16]. The genus *Tuckermanella* contains two foliate species, *T. ciliaris* and *T. fendleri*, with brownish-colored thallus and ellipsoidal ascospores. These species are endemic of North America [8]. Finally, within this clade, the genus *Tuckermannopsis* is included. Species like *T. cilaris*, *T. chlorophylla* and *T. orbata* are distributed in Japan and North America. *Tuckermannopsis americana* presents greenish-brown foliaceuos thallus [17].

Dactylinia clade is constituted by two arctic-alpine species, *D. arctica* and *D. ramulos*a, [8] with fruticulous pale yellow thallus with branched and prominent podetia.

Melaniela clade presents two species, *M. hepatizon* and *M. stygia*, with blackish foliaceous thallus.

And finally, Esslingeriana clade which contains a single specie *Esslingeriana idahoensis* that grows on coniferous branches and trunks of high areas (1,000-5,000 m altitude). Morphologically, they are characterized by a grayish-green thallus and black apothecia and pycnidia at the margins of the lobes [18].

Other authors classify these subclades as genera and reunify all the genera as *Cetraria, Nephromopsis, Dactylina, Melaniela* and *Esslingeriana*, renaming all the species. This is based on their proximity at DNA level [19].

Chemical Composition

Lichens have a varied and complex chemistry. Thin-layer chromatography (TLC) and high-performance liquid chromatography (HPLC) are the most commonly used analytical techniques for the identification and isolation of compounds present in lichens. Lichen metabolites are divided into two groups: primary and secondary. Among primary metabolites, there have been aminoacids, polysaccharides, proteins, vitamins and carotenoids in cytoplasm and cell walls. Most of these compounds are non-specific to lichens, finding them in other organism such as in free-living fungi, algae, and higher plants [20].

Secondary metabolites are mostly organic compounds with polycyclic nature. Among secondary metabolites, there have been identified several compounds which are exclusive of these organisms. Fungus is the main producer of these compounds, but highly influenced by algae that provide fungus with nutrients thanks to its photosynthetic activity. Isolated cultive of fungus produce substances different from those produced by lichen [21]. Moreover, several studies suggest that photobiont, especially cyanobacteria and bacterial microbiomes, are also an important source of active metabolites [4]. Histologically, secondary metabolites are deposited in the cortex or, more commonly, in the medulla and their location in the lichen thallus is probably linked to its biological function. Cortical compounds are regarded as a kind of light filter which is apparently not a function of compounds below the algal layers [22].

Lichen secondary metabolites are derived from three metabolic pathways: 1) acetate-polymalonate pathway, 2) mevalonic acid pathway and 3) shikimic acid pathway. Most lichen compounds including depsids, depsidones and dibenzofurans are synthesized through acetate-polymalonate pathway which uses acetyl-CoA and malonyl-CoA [23]. Carotenoids, steroids and terpenes are synthesized by mevalonic acid route. This pathway involves 3-hydroxy-3-methylglutaryl-CoA (HMG)-CoA synthesis from acetyl-CoA to acetoacetylCoA. 3-hydroxy-3-methylglutaryl-CoA reductase (HMGR) catalyzes HMG-CoA conversion into mevalonic acid [24]. Triterpenes such as zeorine are the most abundant compounds synthesized by this route. The third synthesis pathway of secondary metabolites is shikimic acid pathway. The condensation of phosphoenolpyruvate (PEP) and erythrose 4-phosphate (E4P) takes place forming shikimic acid. The shikimic acid is phosphorylated and another PEP molecule binds to shikimic 3-phosphate acid. After two sequential eliminations of phosphoric acid, the final product corismic acid is synthesized [25, 26]. This pathway is involved in the synthesis of vulpinic acid.

Table 1 shows chemical composition of species belonging to the cetrarioid clade. Figure 1 shows the chemical structure of some major secondary metabolites of lichens of cetrarioid clade. The most common secondary metabolites in lichen species of Cetrarioid clade are caperatic acid, usnic acid, fumarprotocetraric acid and alectoronic acid, among others.

Table 1. Chemical composition of species belonging to the cetrarioid clade (Superscripts refer to specific compounds identified in some lichen species)

Lichen	Secondary metabolites	References
Ahtiana		
A. aurescens (Tuck.) Thell & Rand *A. pallidula* (Riddle) Goward & Thell. *A. sphaerosporella* (Müll. Arg.)	Caperatic acid, lichesterinic acid, protolichesterinic acid, usnic acid	[14, 27]
Allocetraria		
A. ambigua (C. Bab.) Kurok. & M. J. Lai *A. flavonigrescens*[1] (A. Thell & Randlane) *A. globulans* (A. Thell & Randlane) *A. isidiigera*[1] (Kurok. & M. J. Lai) *A. madreporiformis* (Kärnefelt & A. Thell) *A. sinensis* (X. Q. Gao)	Caperatic acid, lichesterinic acid, protolichesterinic acid, secalonic acid A and/or C, usnic acid [1]Fumarprotocetraric acid, protocetraric acid	[9, 28]
Artocetraria		
A. nigricascens (Nyl.) Kärnefelt & A. Thell.	Caperatic acid, norrangiformic acid, rangiformic acid	[15, 29]
Cetraria		
C. aculeata[1] (Schreb.) *C. ericetorum* Opiz *C. islandica*[2] (L.) Ach *C. laevigata*[3] Rass *C. muricata* (Ach.) Eckfeldt *C. nigricascens* (Nyl.) Elenkin *C. obtusata*[4] (Schaer.) v d Boom & Sipman	Lichesterinic acid, protolichesterinic acid [1]Nephrosterinic acid [2]Fumarprotocetraric acid, protocetraric acid [3]Fumarprotocetraric acid [4]Caperatic acid, secalonic acid	[9, 30]
Dactylina		
Dactylina arctica[1] (Richardson) Nyl. *Dactylina ramulosa*[2] (Hook. f.) Tuck	[1]Barbatic acid, gyrophoric acid, physodic acid, usnic acid [2]Physodalic acid, physodic acid, usnic acid	[31, 32]
Esslingeriana		
E. idahoensis (Hook. f.) Tuck	Atranorin	[33]
Flavocetraria		
F. cucullata[1] (Bellardi) Kärnefelt & A. Thell. *F. nivalis*[2] (L.) Kärnefelt & A. Thell	Usnic acid [1]Baeomicesic acid, lichesterinic acid, protolichesterinic acid, salazinic acid, squamatic acid [2]Epifriedelin, friedelin, lupeol	[29, 34, 35, 36]
Cetrariella		
C. commixta[1] (Nyl.) A. Thell & Kärnefelt *C. delisei*[2] (Bory ex Schaer.) Kärnefelt & A. Thell. *C. fastigata*[2] (Delise ex Nyl.) Kärnefelt & A. Thell.	[1]Alectoronic acid, α-collatolic acid [2]Gyrophoric acid, hiascic acid	[11, 33, 37, 38]

Table 1. (Continued)

Lichen	Secondary metabolites	References
Cetreliopsis		
C. endoxanthoides (D. D. Awasthi) Randlane & Saag. *C. laeteflava*[1] (Zahlbr.) Randlane & Saag. *C. papuae*[2] Randlane & Saag. *C. rhytidocarpa*[2] (Mont. & Bosch) Randlane & Saag *C thailandica*[3] Elix & M. J. Lai.	Fumarprotocetraric acid, protocetraric acid, salazinic acid, usnic acid [1]Conquasetic, quaesitic acids [2]Conquasetic, protolichesterinic acid, quasetic acid, succinprotocetráric acid. [3]Conquasetic, protolichesterinic acid quaesitic acid	[8, 29, 37]
Kaemefeltia		
K. merrillii (Du Rietz) A. Thell & Goward.	Lichesterinic acid, protrolichesterinic acid	[37, 39]
Melanelia		
M. hepatizon[1] (Ach.) A. Thell *M. stygia*[2] (L.) Essl.	[1]Atranorin, cloroatranorin, caperatic acid, criptostictic acid, constictic acid, menegazziaic acid, norstictic acid, stictic acid [2]Stenosporic acid, perlatolic acid	[18, 29, 34]
Masonhalea		
M. inermis[1](Nyl.) Lumbsch, Nelsen & A. Thell) *M. richardsonii*[2] (Hook.) Karnef.	[1]Lichesterinic acid, protolichesterinic acid [2]Alectoronic acid	[18, 37]
Nephromopsis		
N. endocrocea[1a](Asah) *N. komarovii*[1b] (Elenkin) J. C. Wei *N. laii* (A. Thell & Randlane) Saag & A. Thell *N. laureri*[1] (Kremp.) Kurok. *N. leucostigma* (Lév.) A. Thell & Randlane *N. morrisonicola*[2] M. J. Lai *N. nephromoides*[2a] (Nyl.) Ahti & Randlane *N. ornata*[1c] (Müll. Arg.) *N. pallescens*[2b](Schaer.) Y. S. Park *N. stracheyi*[1d] (C. Bab.) Müll. Arg.	[1]Lichesterinic acid, protolichesterinic acid, usnic acid. [a]Endocrocin, isonephrosterinic acid, nephrosterinic acid. [b]Constictic ácid, fumarprotocetraric acid, stictic acid [c]Fumarprotocetraric acid, secalonic C acid [d]Anziaic acid, nephromopsic acid, olivetolic acid [2]Lichesterinic acid, protolichesterinic acid [a]caperatic acid [b]alectoronic acid, α-collatolic acid, usnic acid	[18, 29, 40, 41]

Lichen	Secondary metabolites	References
Tuckermanella		
T. pseudoweberi (Essl.). *T. subfendleri* (Essl.) *T. weberi[1]* (Essl.)	Caperatic acid [1]Alectoronic acid	[29]
Tuckermannopsis		
T. americana[2] (Spreng.) Hale *T. chlorophylla[2a]*(Willd.) Hale *T. cilaris[2a]* (Ach.) Gyeln *T. ulophylloides[2]* (Asahina) M. J. Lai	[1]Alectoronic, α-collatolic acid [2]Lichesterinic acid, protolichesterinic acid [a]Olivetolic acid	[18, 29, 40]
Usnocetraria		
U. oakesiana (Tuck.) M. J. Lai & J. C. Wei.	Caperatic acid, lichesterinic acid, protolichesterinic acid, secalonic acid, usnic acid	[37, 42]
Vulpicida		
V. canadiensis (Räsänen) J.-E. Mattsson & M. J. Lai *V. juniperus[1]* (L.) J.-E. Mattsson & M. J. Lai *V. pinastri[1]* (Scop.) J.-E. Mattsson & M. J. Lai.	Pinastric acid, usnic acid, vulpinic acid [1]Zeorin	[12, 13, 37, 43]

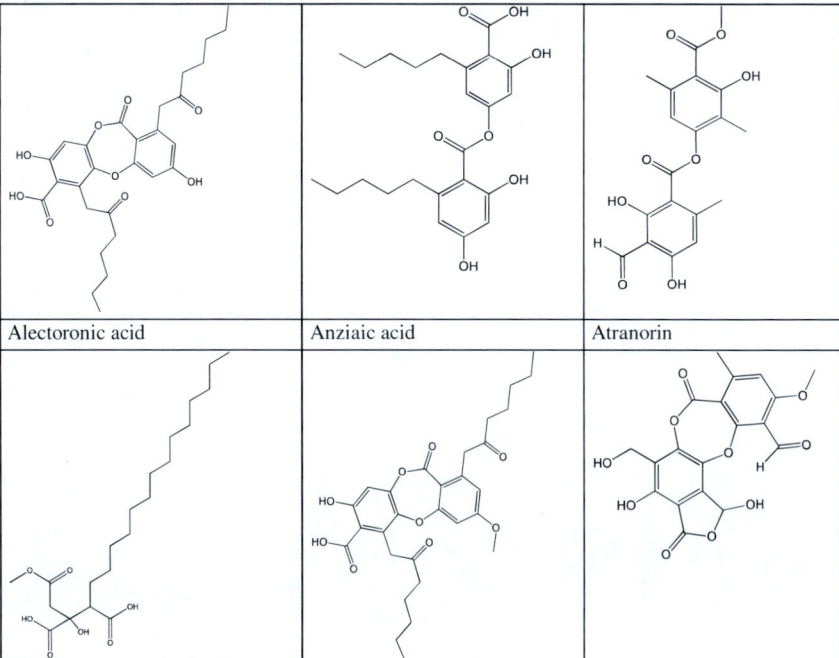

Figure 1. Main compounds from lichens of Cetrarioid Clade.

Caperatic acid	α-Collatolic acid	Constictic acid
Fumarprotocetraric acid	Gyrophoric acid	Lichesterinic acid
Norstictic acid	Olivetolic acid	Physodalic acid
Physodic acid	Pinastric acid	Protocetraric acid
Protolichesterinic acid	Salazinic acid	Secalonic C acid
Stictic acid	Usnic acid	Vulpinic acid

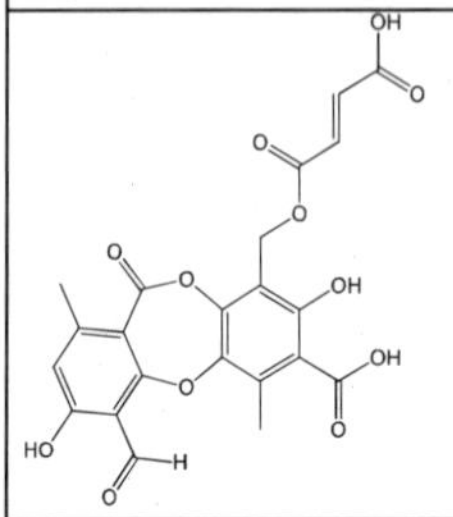

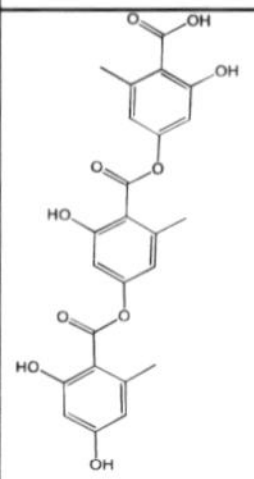

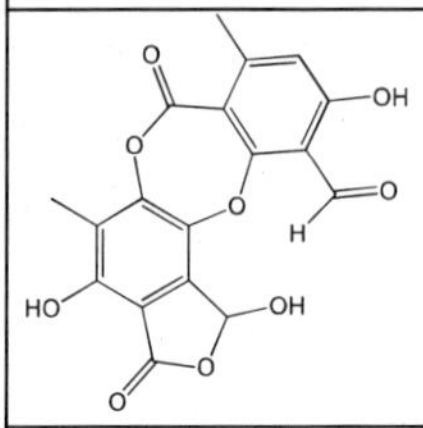

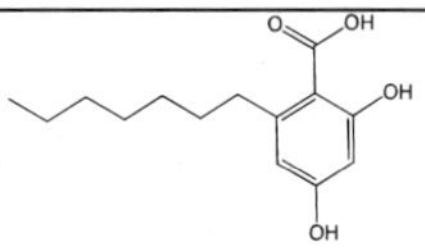

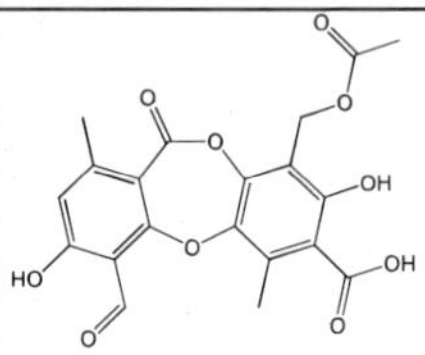

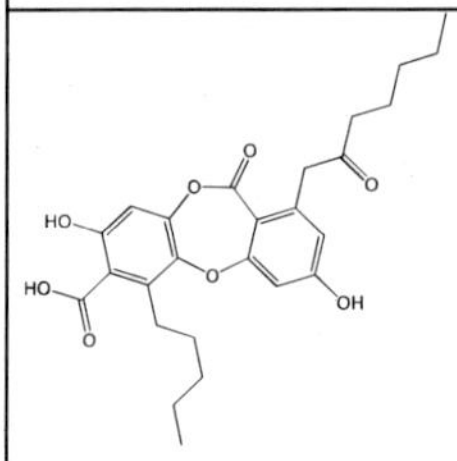

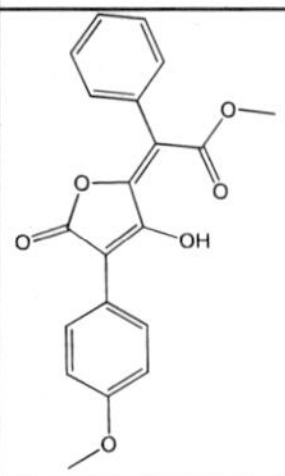

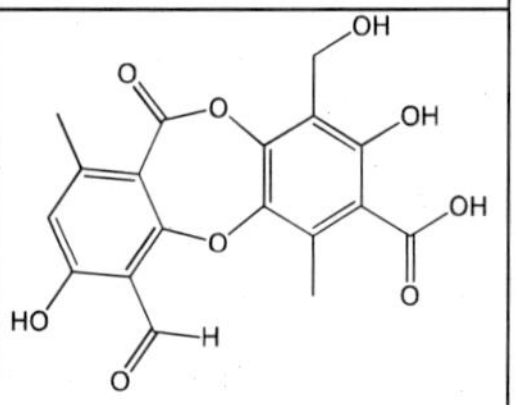

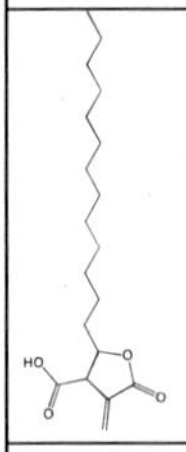

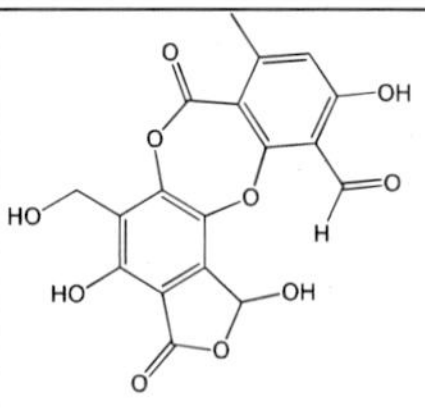

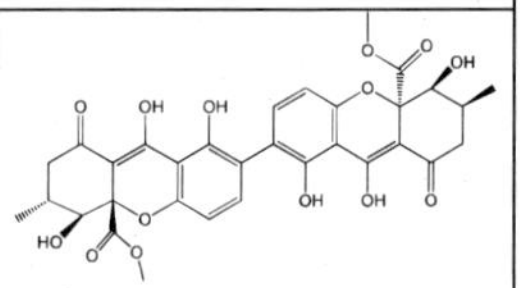

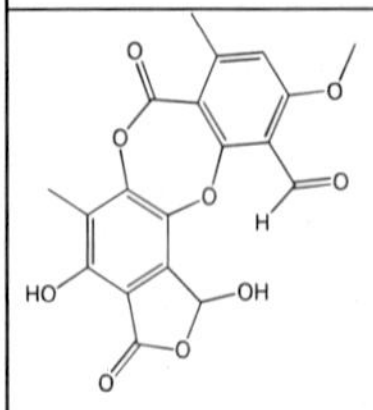

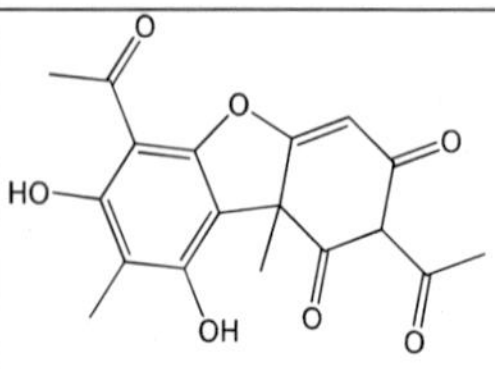

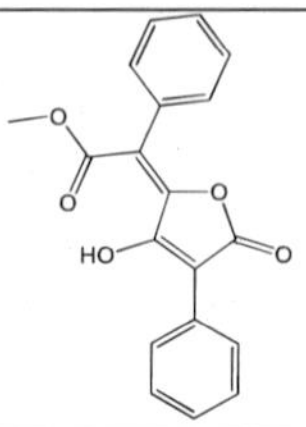

Figure 1 (continued). Main compounds from lichens of Cetrarioid Clade.

PHARMACOLOGICAL INTEREST

Oxidative Stress and Antioxidants: Their Role in Human Diseases

Oxidative stress is defined as an imbalance between reactive oxygen species (ROS) production and the antioxidant system (enzymatic and non-enzymatic).

Reactive oxygen species (ROS) such as hydrogen peroxide (H_2O_2), superoxide anion ($O_2^{.-}$) and hydroxyl radical ($^.OH$) are products of metabolism that act as oxidizing molecules. Particularly, hydrogen peroxide is produced during aerobic respiration and peroxisomal lipid oxidation; it inactives enzymes such as sedoheptulose-bisphosphatase and fructose bisphosphatase. Superoxide anion radical is formed during neurotransmitters autooxidation and in complexes I and III of mitochondrial electron transport chain; this free radical inactivates enzymes which contain iron-sulfur center. Hydroxyl radical is formed by peroxynitrite decomposition and in Haber-weiss and Fenton reactions; this is the most potent free radical [45].

ROS overproduction causes redox homeostasis imbalance, which can damage biological structures such as proteins, DNA and lipids, eventually inducing cell death, mainly due to apoptosis mechanisms.

Oxidative stress has been implicated in the development of many chronic diseases, most associated with age, such as neurodegenerative diseases (i.e., Parkinson's disease and Alzheimer's disease), atherosclerosis, cancer and diabetes [46].

Antioxidants could prevent and even remove oxidative stress through scavenging free radicals excess, metal chelation mechanism and upregulation of antioxidant system. Antioxidants can be classified based on their origin in, endogenous and exogenous and based on their enzymatic and non-enzymatic nature [47].

Endogenous Antioxidants

Endogenous Enzymatic Antioxidants

Catalase (CAT)

This homo-tetrameric enzyme, located in cytosol (20%) and peroxisomes (80%), catalyzes hydrogen peroxide decomposition into oxygen and water.

$$2 H_2O_2 \xrightarrow{\text{Catalase}} O_2 + 2 H_2O$$

Superoxide dismutase (SOD)

There have been identified three diferente Superoxide dismutase enzymes: SOD1 (CuZnSOD), SOD2 (MnSOD), and SOD3 (extracellular, Cu-Zn). These enzymes catalyze the conversion of superoxide anions into oxygen and hydrogen peroxide.

$$O_2^{\bullet-} + O_2^{\bullet-} + 2H^+ \xrightarrow{\text{Superoxide dismutase}} O_2 + 2 H_2O_2$$

Glutathione peroxidase (GPx)

There have been identified two different types of glutathione peroxidases: selenium-dependent glutathione peroxidases [GPX1 which is expressed in cytosol of many tissues, GPX2 which is expressed in epithelium of gastrointestinal tract, GPX3 which is an extracellular enzyme expressed in different tissues such as kidney, prostate, heart, lung, breast and placenta, and GPX4 which is expressed in brain] and selenium-independent glutathione peroxidases which are presented in mitochondria, cytosol and membranes. The enzyme glutathione peroxidase catalyzes the reduction of hydrogen peroxide into water using reduced glutathiones as cofactor.

Glutathione peroxidase

$$H_2O_2 + 2GSH \rightarrow 2H_2O + GSSG$$

Glutathione Reductase (GR)

Glutathione reductase catalyzes reduction of oxidized glutathione (GSSG) into reduced glutathione (GSH) using NADPH cofactor. This enzyme which is located in mitochondria and cytosol participates with glutathione peroxidase in maintaining redox homeostasis.

Glutathione reductase

$$NADPH + H^+ + GSSG \rightarrow NADP^+ + 2GSH$$

Heme Oxygenase-1 (HO-1)

Heme oxygenase-1 catalyzes the oxidative breakdown of heme group to produce biliverdin, ferrous and carbon monoxide. Biliverdin is an efficient scavenger against hydroxyl radical, superoxide anion radical and peroxynitrite.

Heme oxygenase-1

$$\text{Heme group} \rightarrow CO + Fe^{2+} + \text{biliverdin}$$

Endogenous Non-Enzymatic Antioxidants

Glutathione

Glutathione (L-gamma-Glutamyl-L-cysteinyl-glycine) is the major endogenous non-enzymatic antioxidant. It is located in cytosol of cells with a concentration ranging from 1 mM to 10 mM. Glutathione in its reduced state (GSH) scavenge peroxides and diverse xenobiotic compounds through its cysteine thiol group [48].

Apart from glutathione, other endogenous compounds have demonstrated antioxidant properties such as uric acid, bilirubin, melatonin and alpha lipoic acid.

Exogenous Antioxidants

A variety of exogenous compounds with antioxidant properties (of natural origin, semi-synthetic and synthetic) have been identified. It is worth highlighting vitamin E (lipid-soluble antioxidant) which inhibits lipid peroxidation by scavenging lipid peroxides and lipid hydroxyl radicals, and vitamin C (water-soluble antioxidant) which scavenge free radicals (i.e., hydroxyl radical, alkoxyl radical and superoxide radical anion), it also protects against lipid peroxidation and regenerates vitamin E in cell membranes. Other exogenous antioxidants widely investigated are β-carotene (which is a good quencher for singlet oxygen), resveratrol (which acts as metal chelator and modulates antioxidant enzymatic activity), caffeic acid, gallic acid and quercetin [49].

Lichens of Cetrarioid Clade in the Prevention of Oxidative Stress-Related Diseases

This part of the chapter is divided into two sections: antioxidants and prevention of oxidative stress-related diseases. The antioxidant section refers only to *in vitro* technical studies of the antioxidant properties of lichen extracts and active compounds identified in species of the cetrarioid clade. The section prevention of oxidative stress-related diseases is subdivided into pathologies and/or activities: neurodegenerative diseases, immune system, diabetes, atherosclerosis, genotoxicity and cancer, respiratory diseases and antimicrobial and it refers to the protective and preventive role of lichen extracts and their isolated compounds based on their antioxidant properties.

In this chapter, it should be noted that some of the secondary metabolites with antioxidant properties have not been isolated in species of the cetrarioid clade but there have been identified in them.

1. Antioxidants.

Antioxidants are molecules that inhibit ROS action or slow down biomolecules oxidation. The use of exogenous antioxidants is considered the most promising strategy to cope with oxidative stress.

There is a growing research interest in searching natural compounds with potential antioxidant properties. Previous works have demonstrated that lichen secondary metabolites, mainly those with phenolic compounds, are promising antioxidant compounds.

Several *in vitro* techniques have been used to evaluate the antioxidant activity of cetrarioid lichen extracts and its isolated secondary metabolites. These techniques are based in the ability of lichens to single electron transfer (SET), to hydrogen atom transfer (HAT) and to chelate metals.

The most commonly used antioxidant technique in lichens has been 2,2'-diphenyl-1-picrylhydrazyl (DPPH) method. DPPH is a relatively stable free radical in solution. It has intense purple coloration, which turns pale yellow in presence of a compound capable of donating electrons that stabilize its molecule [50]. The lower absorbance (lower intensity of purple color), the greater antioxidant activity of the sample. This technique has the advantage of being very simple and quick, but it does not allow to estimate antioxidant action over time, and it has poor reactivity. DPPH assay has been used to investigate the antioxidant activity of lichens of the genus *Cetraria*, *Melanelia*, *Vulpicida*, *Flavocetraria* and *Nephromopsis*. Within *Cetraria* genus, the antioxidant activity of the species *C. laevigata* and *C. islandica* has been studied. *Cetraria laevigata* has shown an IC_{50} value in DPPH radical scavenging activity of 55 µg/mL [43, 51]. On the other hand, the antioxidant activity of *C. islandica* using DPPH assay varies from 678.38 µg/mL to 1183 µg/mL depending on the study [43, 52]. For *Melanelia* genus, *M. subaurifera* was more potent antioxidant than *M. fuliginosa* as evidenced in IC_{50} values for DPPH (165 µg/mL *versus* 266 µg/mL) [53]. Moreover, the species of *Vulpicida* genus, *V. canadiensis* and *V. pinastri*, exhibited significant inhibition in DPPH free radical formation with IC_{50} values of 99 and 75 µg/mL, respectively [43, 54]. Furthermore, the specie *F. cucullata* exhibited potent DPPH radical scavenging activity which was attributed to cuculloquinone [55]. Finally, species *Nephromopsis nephromoides* and *N. ornata* possessed moderate antioxidant activity as revealed IC_{50} values of 110 and 254 µmol/L [51].

Apart from the antioxidant activity of lichen extracts based on DPPH method, the activity has been evaluated for secondary metabolites identified in different lichen species of cetrarioid clade. Hence, alectoronic acid, atranorin and α-collatolic acid at a concentration of 100 μg/mL showed IC_{50} values of 23.62 μg/mL, 34.26 μg/mL and 20 μg/mL, respectively, for DPPH radicals [56]. Moreover, anziaic acid (IC_{50} value of 121.52 μg/mL), fumarprotocetraric acid (IC_{50} value of 228.46 μg/mL), gyrophoric acid (IC_{50} value of 105.75 μg/mL), norstictic acid (IC_{50} value of 102.65 μg/mL) and salazinic acid (IC_{50} value of 91.57 μg/mL) exhibited DPPH scavenging activity[53,57-60].Usnic acid was found as potent DPPH (IC_{50} value 49.50 μg/mL) [61]. Furthermore, other compounds such as zeorin did not show antioxidant activity using DPPH assay [62].

Another technique commonly used to evaluate antioxidant activity in lichens is Oxygen radical absorbance capacity (ORAC) assay. This method measures antioxidant properties of compounds based on its capacity to neutralize peroxyl radicals generated by thermal decomposition of the azo compound 2,2'-azobis (2-amidinopropane)-dihydrochloride] (APPH) [63]. This hydrogen atom transfer method has the advantage to detect both hydrophilic and hydrophobic antioxidants. However, it is necessary to control various factors (i.e., pH, temperature) during experiment to be reproducible, and fluorescence quenching is very sensitive [64]. *Cetraria islandica* had an ORAC value of 3.06 TE/mg dry extract [43], *Masonhalea richardsonii* of 829.1 TE/μg dry extract [65] and *Vulpicida canadiensis* of 0.77 TE/mg dry extract. Moreover, the secondary metabolite stictic acid showed an ORAC value of 2.32 μmol TE/mg.

Ferric reducing antioxidant power (FRAP) assay has been used to evaluate the antioxidant activity of secondary metabolites found in cetrarioid lichens. This single electron transfer method is based on the ability of antioxidants to reduce ferric ion into ferrous ion. Subsequently, ferrous ion may interact with 2,4,6-tris(2-pyridyl)-s-triazine (TPTZ) producing a colored complex with maximum absorbance at 593 nm [66]. This is a simple and quick method with the limitation of no exact reaction time of samples. Hence, salazinic acid showed a FRAP value of 37.18 mmol Fe (II)/g sample [67]. Moreover, zeorin, at a concentration of 50 μg/ml, had a FRAP value of 63.7% [68].

Furthermore, several works aimed to assess superoxide radical and hydroxyl radical scavenging activities of lichen extracts and secondary metabolites. Hence, the value of the superoxide scavenging activity for *Cetraria islandica* was 792.48 µg/mL [52]. The compound salazinic acid inhibited hydroxyl radical with IC_{50} value of 193.36 µg/mL and superoxide radical with IC_{50} value of 119.1-294.07 µg/mL [67, 69]. Moreover, fumarprotocetraric acid, norstictic acid and stictic acid showed superoxide anion scavenging activity with values of IC_{50} 105.75 µg/ml, 133.46 µg/ml and 104.34 µg/ml, respectively [57, 59, 70]. Furthermore, zeorin at a concentration of 50 µg/mL inhibited free radicals with a similar potency to that of reference compounds such as Trolox, BHA and BHT (Free radical scavenging value of 27.8%, Superoxide anion scavenging of 39.6%, hydroxyl radical scavenging of 58.8% and Hydrogen peroxide scavenging of 48.6%) [68]. In addition, salazinic acid showed metal chelating activity with a value of 6.79 mg EDTA eq/g sample, thus avoiding the generation of reactive oxygen species as well as a value of 117.08 mg ascorbic acid eq/g sample in phosphomolybdenum method [67].

The Trolox equivalent antioxidant capacity (TEAC) assay was used to determine the antioxidant capacity of *Cetrariella fastigata*. This method revealed a TEAC value of 7.8 mM [71, 72].

Finally, there is a strong correlation between antioxidant activity and phenols compounds. Folin-Ciocalteu method is commonly applied to evaluate phenolic content. Phenolic compounds react with Folin-Ciocalteu reagent which is composed by phosphotungstic-phosphomolybdenum. The reaction forms a blue chromophore complex that can be quantified by visible-light spectrophotometry. Standard curve is usually made with gallic acid, tannic acid, catechin and pyrogallol can also be used as references [73, 74]. Phenolic content determination in *Cetraria islandica* and *Vulpicida canadiensis* show TLC values of 57.34 and 34.93 µg GA/mg respectively [43].

Prevention of Oxidative Stress-Related Diseases

Neurodegenerative Diseases

Neurodegenerative diseases are chronic and progressive disorders underlying cause of dendrites injury, demyelination and neuronal death and as consequence, cognitive impairment (dementia) and/or loss of motor skills [75]. Oxidative stress is involved in the pathophysiology of many neurodegenerative diseases. Brain is especially vulnerable to oxidative damage due to its high content of oxidizable substrates such as catecholamines and polyunsaturated fatty acids (PUFAs), its high oxygen consumption (20%), low content of antioxidant enzymes and glutathione and, accumulation of transition metals in certain brain regions [76]. Hence, post-mortem brain analysis of patients suffered from Alzheimer's disease have identified high levels of lipid peroxidation products (malondialdehyde and 4-hydroxynonenal), protein nitration content and DNA oxidation (8-OHdG level increased) [77]. Amyloid beta peptide (Aβ) promotes hydrogen peroxide production by reducing metal copper and iron ions [78]. On the other hand, studies on brain of patients with Parkinson's disease have shown that carbonylated and nitrosylated proteins are found in high amounts and, catalase and peroxidase enzymatic activity and glutathione content are low [79]. Hydrogen peroxide is produced as a by-product of dopamine oxidation by monoamine oxidases A and B [80].

Methanol extracts of *Vulpicida canadensis* and *Cetraria islandica* demonstrated neuroprotective activity against hydrogen peroxide-induced oxidative stress in the human astrocytoma U373-MG cell line. Particularly, *Vulpicida canadensis* at a concentration of 25 µg/mL and *Cetraria islandica* at a concentration of 10 µg/mL increased significantly cell viability and improved cell morphology. Moreover, these lichen extracts reduced intracellular ROS production, lipid peroxidation levels and caspase-3 activity as well as they ameliorated the gluthatione system [43].

The secondary metabolite fumarprotocetraric resulted to be a promising neuroprotective agent against hydrogen peroxide-induced oxidative stress in the human U373 MG astrocytic and SH-SY5Y neuronal cell lines. Fumarprotocetraric at a concentration of 1 µg/ml (for U373 MG cell line) and 25 µg/ml (for SH-SY5Y cell line) reduced intracellular ROS generation, lipid peroxidation and GSH depletion as well as it increased antioxidant enzymes expression (catalase, superoxide dismutase 1 and hemeoxigenase-1). Moreover, this compound increased mitochondria membrane potential and mitochondrial calcium, thus exerting a protective effect of mitochondrial-targeted. Furthermore, since oxidative stress play a key role in apoptosis, it was demonstrated that pre-treatments with fumarprotocetraric increased Bcl-2 protein expression and decreased Bax protein expression. Thus, cytoprotective action of fumarprotocetraric seems to be related, at least in part, with its ability to activate Nrf2 signaling pathway [81].

In another study, Toledo Marante et al. evaluated the protective effect of atranorin, presented in various lichen species of cetrarioid clade, on lipid peroxidation in an oxidative stress model of H_2O_2/Fe^{2+} induced. As it was evidenced, atranorin (100 and 250 µM) reduced malonaldehyde and 4-hydroxyalkenals content in the brains of Sprague–Dawley rats [82].

Furthermore, the secondary metabolite stictic acid increased U373MG cell viability and reduced intracellular ROS production at a concentration of 5 µg/mL [69].

Finally, usnic acid showed moderate neuroprotective effect against redox impairment hydrogen peroxide mediated cytotoxicity in the human astrocytoma U373 MG cell line and in the human SH-SY5Y neuroblastoma cell line by reducing ROS production, recovering glutathione system and decreasing lipid peroxidation levels [83].

Immune System

The immune system is conditioned by redox balance since ROS overproduction alters structural integrity and function of immune cells [84].

Cernescu et al. investigated the effect of *Cetraria islandica* extract in Wistar rats with levamisole-induced granuloma. This study demonstrated that the intraperitoneally administration of *Cetraria islandica* extract (21.56 mg/kbw/day) reduced lipid peroxidation levels as evidenced in malondialdehyde content and increased glutathione levels [85]. Moreover, these researchers observed a synergistic effect between the antioxidant action of the lichen extract and magnesium. In another study, Kotan et al. demonstrated that methanol extract of *Cetraria islandica* at concentrations of 5 µg/mL and 10 µg/mL increased the enzymatic activities of superoxide dismutase and glutathione peroxidase and reduced lipid peroxidation in human lymphocytes [86].

Usnic acid (1 and 5 µg/ml) showed antioxidant activity in cultured human peripheral blood cells as evidenced in total antioxidant capacity level [87]. Moreover, the lichen compound physodic acid (0.5-10 mg/L) increased total antioxidant capacity in human lymphocytes [88]. Furthermore, physodic acid at lower concentrations than 50 mg/L showed antioxidant properties in cultured human amnion fibroblasts (HAFs) being of interest to protect against oxidative stress [89]. Moreover, this lichen compound increased ROS production and it caused mitochondrial membrane potential disturbances in isolated rat thymocytes [90].

Diabetes

Oxidative stress is involved in the pathogenesis of the metabolic disorder diabetes. The auto-oxidation of glucose and non-enzymatic glycation of proteins produce ROS (i.e., hydroxyl radicals), leading to insulin resistance and macromolecules oxidation (proteins such as glycated hemoglobin, lipids such as lipid hyperperoxides and DNA) [91].

The aqueous extract of *Cetraria islandica* at concentrations of 5 and 10 mg/mL increased the activity of the antioxidant enzymes superoxide dismutase, catalase and glutathione peroxidase in the erythrocytes of patients with diabetes. Moreover, this lichen extract reduced lipid peroxidation and DNA damage [92]. On the other, Colak et al. demonstrated that the aqueous extract of *Cetraria islandica* (250 and 500 mg/kg/day for 2 weeks), based on its antioxidant properties, could be just beneficial in streptozotocin (STZ)-induced diabetes in rats in early stages by increasing total antioxidant capacity [93, 94].

Atherosclerosis

Oxidative stress also contributes to atherosclerosis. Overproduction of ROS modify vascular tone and cause LDL-cholesterol oxidation [95].

Vulpinic acid has demonstrated to be a promising agent to prevent oxidative stress associated with atherosclerosis. This compound at a concentration of 15 μM significantly reduced intracellular ROS production and cell damage in a model of induced hydrogen peroxide in the human umbilical vein endothelial cells (HUVECs) [96].

Genotoxicity and Cancer

The role of oxidative stress in cancer is controversial since ROS overproduction could promote or could suppress tumor cell survival and progression via gene mutations induction or via signal transduction and transcription factors [97, 98]. Cancer promotion or cancer suppression depend on stress intensity and type of ROS [99].

In a recent work, it has been shown that treatment time with usnic acid determined antioxidant-prooxidant activity. Hence, 24 h treatments with usnic acid increased ROS generation and reduced GSH content in the human liver cancer HepG2 cell line whereas 6 h treatments with usnic acid induced Nrf-2 signaling pathway [100]. Moreover, usnic acid exerted anti-cancer action by increasing ROS levels and causing cellular mitochondrial dysfunction in a concentration dependent manner in the human hepatoblastoma HepG2 cells [101].

Usnic acid combined with bleomycin inhibited tumor growth in ascitic H22-bearing mice among other mechanisms through its antioxidant activity, reducing lipid peroxidation and increasing superoxide dismutase activity [102]. Moreover, usnic acid (50 and 100 mg/kg) had genotoxic effect in mice as shown in an increase of malondialdehyde content and DNA damage in liver and kidneys [103].

The compound gyrophoric acid protected against cervix carcinoma by increasing ROS generation leading to DNA damage and induced apoptosis via the AKT/p38 MAPK/ERK1/2 signaling pathway in the human cervical carcinoma HeLa cell line [104].

Finally, physodic acid resulted to be an *in vitro* antitumor agent in the human primary gliobastoma U87MG cell line (at a concentration of 410.72 mg/mL) and in primary rat cerebral cortex (PRCC) cells (at a concentration of 698.19 mg/L) by causing oxidative DNA damage [105].

Respiratory Diseases

The respiratory system is particularly susceptible to oxidative stress because high content in oxygen in lungs and presence of exogenous pollutants (i.e., cigarette smoke, ozone and nitrogen dioxide) which could induce ROS generation. Because of redox impairment in respiratory system, there is an increase in mucus secretion, lung inflammation, bronchoconstriction and edema [106].

Pretreatments with usnic acid (50 or 100 mg/kg body weight, p.o.) reduced myeloperoxidase activity, malondialdehyde content and hydrogen peroxide level on Lipopolysaccharides-induced acute lung injury in mice. Moreover, this lichen compound exerted protective effect by increasing superoxide dismutase activity and GSH levels [107].

Antimicrobial

Bacteria have antioxidant defense system to detoxy ROS generation; however, when this production exceeds the antioxidant capacity of bacteria, it favors antimicrobial activity. ROS-inspired antimicrobials act through two different mechanism of action: as oxidative stress inductors pharmacophores-target and by compromising antioxidant defense system [108].

Usnic acid at a concentration of 4 µg/mL inhibited azole-resistant *Candida albicans* growth by promoting reactive oxygen species and reactive nitrogen species generation [109]. Moreover, usnic acid combined with norfloxacin also inhibited multidrug-resistant *Staphylococcus aureus* infections by inducing oxidative stress, as evidence in membrane damage and peptidoglycan and fatty acid biosynthesis inhibition [110].

CONCLUSION

Lichens and their secondary metabolites have potential antioxidant activity, showing clinical interest for the prevention and treatment of pathologies associated with oxidative stress. It should be noted that within the cetrarioid clade, many lichen species have not been investigated. Indeed, many secondary compounds which have shown *in vitro* antioxidant properties, have not been yet tested in cell lines and animal models. Future research should be aimed at deepening the mechanism of action of secondary metabolites and to perform clinical trials.

REFERENCES

[1] Aschenbrenner, Ines A., Tomislav Cernava, Gabriele Berg, and Martin Grube. "Understanding Microbial Multi-Species Symbioses." *Frontiers in Microbiology* 7, (2016):180. https://doi.org/10.3389/fmicb.2016.00180.

[2] Cardinale, Massimiliano, João Vieira De Castro, Henry Muller, Gabriele Berg, and Martin Grube. "In Situ Analysis of the Bacterial Community Associated with the Reindeer Lichen *Cladonia Arbuscula* Reveals Predominance of Alphaproteobacteria." *FEMS Microbiology Ecology* 66, no. 1 (2008): 63–71. https://doi.org/10.1111/j.1574-6941.2008.00546.x.

[3] Padhi, Srichandan, Devaranjan Das, Suraj Panja, and Kumananda Tayung. "Molecular Characterization and Antimicrobial Activity of an Endolichenic Fungus, Aspergillus Sp. Isolated from Parmelia Caperata of Similipal Biosphere Reserve, India." *Interdisciplinary Sciences: Computational Life Sciences* 9, no. 2 (2016): 237–46. https://doi.org/10.1007/s12539-016-0146-y.

[4] Calcott, Mark J., David F. Ackerley, Allison Knight, Robert A. Keyzers, and Jeremy G. Owen. "Secondary Metabolism in the Lichen Symbiosis." *Chemical Society Reviews* 47, no. 5 (2018): 1730–60. https://doi.org/10.1039/c7cs00431a.

[5] Zambare, Vasudeo P., and Lew P. Christopher. "Biopharmaceutical Potential of Lichens." *Pharmaceutical Biology* 50, no. 6 (March 2012): 778–98. https://doi.org/10.3109/13880209.2011.633089.

[6] Divakar, Pradeep K., and D. K. Upreti. *Parmelioid Lichens in India: (a Revisionary Study)*. Dehra Dun: Bishen Singh Mahendra Pal Singh, 2005.

[7] Büdel, B., and C. Scheidegger. "Thallus Morphology and Anatomy." *Lichen Biology*, 40–68. Cambridge University Press. https://doi.org/10.1017/cbo9780511790478.005.

[8] Thell, Arne, Ana Crespo, Pradeep K. Divakar, Ingvar Kärnefelt, Steven D. Leavitt, H. Thorsten Lumbsch, and Mark R. D. Seaward. "A Review of the Lichen Family Parmeliaceae - History, Phylogeny and Current Taxonomy." *Nordic Journal of Botany* 30, no. 6 (2012): 641–64. https://doi.org/10.1111/j.1756-1051.2012.00008.x.

[9] Rai, Himanshu, and Dalip K. Upreti. *Terricolous Lichens in India Volume 2: Morphotaxonomic Studies*. Springer New York, 2014.

[10] Van Den Boom, P. P. G. and H. J. M. Sipman. "*Cetraria Obtusata* Comb. Et Stat. Nov., An Overlooked Lichen Species From the Central Alps." *The Lichenologist* 26, no. 2 (1994): 105–12. https://doi.org/10.1006/lich.1994.1010.

[11] Szczepańska, Katarzyna, and Maria Kossowska. "*Cetrariella Commixta* and the Genus *Melanelia* (Parmeliaceae, Ascomycota) in Poland." *Herzogia* 30, no. 1 (2017): 272–88. https://doi.org/10.13158/heia.30.1.2017.272.

[12] Saag, L., K. Mark, A. Saag, and T. Randlane. "Species Delimitation in the Lichenized Fungal Genus *Vulpicida* (Parmeliaceae, Ascomycota) Using Gene Concatenation and Coalescent-Based Species Tree Approaches." *American Journal of Botany* 101, no. 12 (January 2014): 2169–82. https://doi.org/10.3732/ajb.1400439.

[13] Mattsson, Jan-Erik. "A Monograph of the Genus *Vulpicida* (Parmeliaceae, Ascomycetes)." *Nordic Journal of Botany* 13, no. 4 (1993): 472–72. https://doi.org/10.1111/j.1756-1051.1993.tb00084.x.

[14] Thell, Arne, Trevor Goward, Tiina Randlane, E. I. Kärnefelt, Andres Saag, and E. I. Karnefelt. "A Revision of the North American Lichen Genus *Ahtiana* (Parmeliaceae)." *The Bryologist* 98, no. 4 (1995): 596. https://doi.org/10.2307/3243591.

[15] Brodo, Irwin M., Sylvia Duran Sharnoff, Stephen Sharnoff, and Susan Laurie-Bourque. *Keys to Lichens of North America*: Revised and Expanded. New Haven. Yale University Press, 2016.

[16] Thell, Arne, Tiina Randlane, Andres Saag, and Ingvar Kärnefelt. "A New Circumscription of the Lichen Genus *Nephromopsis* (Parmeliaceae, Lichenized Ascomycetes)." *Mycological Progress* 4, no. 4 (2005): 303–16. https://doi.org/10.1007/s11557-006-0135-3.

[17] Bisby F. A., Y. R. Roskov, T. M. Orrell, D. Nicolson, L. E. Paglinawan, N. Bailly, P. M. Kirk, T. Bourgoin, G. Baillargeon and D. Ouvrard. "Species 2000 & ITIS Catalogue of Life: 2011 Annual Checklist." *Species 2000: Reading, UK.* (2012) http://www.catalogueoflife.org/annual-checklist/2011/search/all/key/tuckermannopsis+americana/match/1.

[18] Hale, Mason E., and Mariette Cole. *Lichens of California.* Berkeley. University of California Press, 1988.

[19] Divakar, Pradeep K., Ana Crespo, Ekaphan Kraichak, Steven D. Leavitt, Garima Singh, Imke Schmitt, and H. Thorsten Lumbsch. "Using a Temporal Phylogenetic Method to Harmonize Family- and Genus-Level Classification in the Largest Clade of Lichen-Forming Fungi." *Fungal Diversity* 84, no. 1 (November 2017): 101–17. https://doi.org/10.1007/s13225-017-0379-z.

[20] Ranković Branislav. *Lichen Secondary Metabolites: Bioactive Properties and Pharmaceutical Potential.* Springer International Publishing, 2019.

[21] Brunauer, Georg, Armin Hager, Martin Grube, Roman Türk, and Elfie Stocker-Wörgötter. "Alterations in Secondary Metabolism of Aposymbiotically Grown Mycobionts of *Xanthoria Elegans* and Cultured Resynthesis Stages." *Plant Physiology and Biochemistry* 45, no. 2 (2007): 146–51. https://doi.org/10.1016/j.plaphy007.01. 004.

[22] Marqués, J. "*A framework for assessing the vulnerability of schist surfaces to lichen-induced weathering in the Upper Douro region (NE Portugal).*" Universidade de Porto. (2013).

[23] Fernández Moriano, Carlos. "*Estudio con criterios filogenéticos del potencial neuroprotector de líquenes parmeliáceos: mecanismos de acción de sus metabolitos secundarios* [Study with phylogenetic criteria of the neuroprotective potential of parmeliaceous lichens: mechanisms of action of their secondary metabolites]." (2017).

[24] Buhaescu, Irina, and Hassane Izzedine. "Mevalonate Pathway: A Review of Clinical and Therapeutical Implications." *Clinical Biochemistry* 40, no. 9-10 (2007): 575–84. https://doi.org/10.1016/j.clinbiochem.2007.03.016.

[25] Nadal, Brice, Sophie A.-L. Thetiot-Laurent, Serge Pin, Jean-Philippe Renault, Damien Cressier, Ghassoub Rima, Antoine Le Roux, Stéphane Meunier, Alain Wagner, and Claude Lion. "Synthesis and Antioxidant Properties of Pulvinic Acids Analogues." *Bioorganic & Medicinal Chemistry* 18, no. 22 (2010): 7931–39. https://doi.org/10.1016/j.bmc.2010.09.037.

[26] Hansson, David. "*Structure and Biosynthesis of Fungal Secondary Metabolites. Studies of the Root Rot Pathogen Heterobasidion annosum s.l. and the Biocontrol Fungus Phlebiopsis giganteara.*" (2013).

[27] Goward, Trevor. "Ahtiana, a New Lichen Genus in the Parmeliaceae" *The Bryologist* 88, no. 4 (1985):367-71. https://doi.org/10.2307/3242678.

[28] Wang, Rui-Fang, Xin-Li Wei, and Jiang-Chun Wei. "The Genus Allocetraria (Parmeliaceae) in China." *Mycotaxon* 130, no. 2 (September 2015): 577–91. https://doi.org/10.5248/130.577.

[29] Xu, Maonian, Starri Heidmarsson, Elin Soffia Olafsdottir, Rosa Buonfiglio, Thierry Kogej, and Sesselja Omarsdottir. "Secondary Metabolites from Cetrarioid Lichens: Chemotaxonomy, Biological Activities and Pharmaceutical Potential." *Phytomedicine* 23, no. 5 (2016): 441–59. https://doi.org/10.1016/j.phymed.2016.02.012.

[30] Ingolfsdottir, K., W. Breu, S. Huneck, G. A. Gudjonsdottir, B. Müller-Jakic and H. Wagner. "In vitro inhibition of 5-lipoxygenase by protolichesterinic acid from *Cetraria islandica*." *Phytomedicine* 1, no. 3 (December 1994): 187-91.

[31] Thomson, J. W., and C. D. Bird. "The Lichen Genus *Dactylina* in North America." *Canadian Journal of Botany* 56, no. 14 (1978): 1602–24. https://doi.org/10.1139/b78-190.

[32] Stocker-Worgotter, Elfie and John A. Elix. "Morphogenetic strategies and induction of secondary metabolite biosynthesis in cultured lichen-

forming Ascomycota, as exemplified by Cladia retipora (Labill.) Nyl. and Dactylina arctica (Richards) Nyl." *Symbiosis* 41, (2006): 9-20.

[33] Øvstedal, D., T. Tønsberg, and A. Elvebakk. "The Lichen Flora of Svalbard." *Sommerfeltia* 33, no. 1 (January 2009): 1–393. https://doi.org/10.2478/v10208-011-0013-5.

[34] Mishra, Gaurav K. and Dalip K. Upreti "Altitudinal distribution of cetrarioid lichens in Govind Wildlife Sanctuary, Garhwal Himalaya, Uttarakhand, India." *Geophytology* 45, no.1 (May 2015): 9-19. ISSN 0376-5561.

[35] Joly, D., L. Nilsen, T. Brossard, and J. W. Bjerke. "Spatial Trends in Usnic Acid Concentrations of the Lichen *Flavocetraria Nivalis* along Local Climatic Gradients in the Arctic (Kongsfjorden, Svalbard)." *Polar Biology* 27, no. 7 (January 2004): 409–17. https://doi.org/10.1007/s00300-004-0590-8.

[36] Nguyen, Thanh Thi, Somy Yoon, Yi Yang, Ho-Bin Lee, Soonok Oh, Min-Hye Jeong, Jong-Jin Kim, et al. "Lichen Secondary Metabolites in *Flavocetraria Cucullata* Exhibit Anti-Cancer Effects on Human Cancer Cells through the Induction of Apoptosis and Suppression of Tumorigenic Potentials." *PLoS ONE* 9, no. 10 (2014). https://doi.org/10.1371/journal.pone.0111575.

[37] Thell, Arne, Filip Högnabba, John A. Elix, Tassilo Feuerer, Ingvar Kärnefelt, Leena Myllys, Tiina Randlane, et al. "Phylogeny of the Cetrarioid Core (Parmeliaceae) Based on Five Genetic Markers." *The Lichenologist* 41, no. 5 (June 2009): 489–511. https://doi.org/10.1017/s0024282909990090.

[38] Seriña, Estela, Rosario Arroyo, Esteban Manrique, Leopoldo G. Sancho, and Estela Serina. "Lichen Substances and Their Intraspecifie Variability within Eleven Umbilicaria Species in Spain." *The Bryologist* 99, no. 3 (1996): 335. https://doi.org/10.2307/3244307.

[39] Brodo, Irwin M., Sylvia Duran Sharnoff, and Stephen Sharnoff. "*Lichens of North America*." New Haven, Yale University Press, 2001.

[40] Shrestha, Gajendra, Jocelyn Raphael, Steven D. Leavitt, and Larry L. St. Clair. "In Vitroevaluation of the Antibacterial Activity of Extracts from 34 Species of North American Lichens." *Pharmaceutical Biology* 52, no. 10 (2014): 1262–66. https://doi.org/10.3109/13880209.2014.889175.

[41] Wang, Yi, Changan Geng, Xiaolong Yuan, Mei Hua, Fenghua Tian, and Changtian Li. "Identification of a Putative Polyketide Synthase Gene Involved in Usnic Acid Biosynthesis in the Lichen *Nephromopsis Pallescens.*" *Plos One* 13, no. 7 (2018). https://doi.org/10.1371/journal.pone.0199110.

[42] Klepsland, J. T. and E. Timdal "*Usnocetraria oakesiana* (Parmeliaceae) new to Northern Europe." *Graphis Scripta* 22, (2010): 14–17. ISSN 0901-7593.

[43] Fernández-Moriano, Carlos, Pradeep Kumar Divakar, Ana Crespo, and M. Pilar Gómez-Serranillos. "Neuroprotective Activity and Cytotoxic Potential of Two Parmeliaceae Lichens: Identification of Active Compounds." *Phytomedicine* 22, no. 9 (2015): 847–55. https://doi.org/10.1016/j.phymed.2015.06.005.

[44] "*PubChem.*" National Center for Biotechnology Information. PubChem Compound Database. U.S. National Library of Medicine. Accessed January 2020. https://pubchem.ncbi.nlm.nih.gov/.

[45] Schieber, Michael, and Navdeep S. Chandel. "ROS Function in Redox Signaling and Oxidative Stress." *Current Biology* 24, no. 10 (2014). https://doi.org/10.1016/j.cub.2014.03.034.

[46] Tan, Bee Ling, Mohd Esa Norhaizan, Winnie-Pui-Pui Liew, and Heshu Sulaiman Rahman. "Antioxidant and Oxidative Stress: A Mutual Interplay in Age-Related Diseases." *Frontiers in Pharmacology* 9 (2018). https://doi.org/10.3389/fphar.2018.01162.

[47] Birben, Esra, Umit Murat Sahiner, Cansin Sackesen, Serpil Erzurum, and Omer Kalayci. "Oxidative Stress and Antioxidant Defense." *World Allergy Organization Journal* 5, no. 1 (2012): 9–19. https://doi.org/10.1097/wox.0b013e3182439613.

[48] Forman, Henry Jay, Hongqiao Zhang, and Alessandra Rinna. "Glutathione: Overview of Its Protective Roles, Measurement, and

Biosynthesis." *Molecular Aspects of Medicine* 30, no. 1-2 (2009): 1–12. https://doi.org/10.1016/j.mam.2008.08.006.

[49] Bouayed, Jaouad, and Torsten Bohn. "Exogenous Antioxidants—Double-Edged Swords in Cellular Redox State: Health Beneficial Effects at Physiologic Doses versus Deleterious Effects at High Doses." *Oxidative Medicine and Cellular Longevity* 3, no. 4 (2010): 228–37. https://doi.org/10.4161/oxim.3.4.12858.

[50] Molyneux, Philip "The use of the stable radical Diphenylpicrylhydrazyl (DPPH) for estimating antioxidant activity." *Songklanakarin Journal of Science and Technology* 26, no.2 (November 2003).

[51] Hara, Kojiro, Marie Endo, Hiroko Kawakami, Masashi Komine and Yoshikazu Yamamoto. "Anti-oxidation activity of ethanol extracts from natural thalli of Lichens" *Mycosystema* 30, no.6 (November 2011): 950-54.

[52] Grujičić, Darko, Ivana Stošić, Marijana Kosanić, Tatjana Stanojković, Branislav Ranković, and Olivera Milošević-Djordjević. "Evaluation of in Vitro Antioxidant, Antimicrobial, Genotoxic and Anticancer Activities of Lichen *Cetraria Islandica*." *Cytotechnology* 66, no. 5 (April 2014): 803–13. https://doi.org/10.1007/s10616-013-9629-4.

[53] Ristić, Svetlana, Branislav Ranković, Marijana Kosanić, Tatjana Stanojković, Slaviša Stamenković, Perica Vasiljević, Ivana Manojlović, and Nedeljko Manojlović. "Phytochemical Study and Antioxidant, Antimicrobial and Anticancer Activities of *Melanelia Subaurifera* and *Melanelia Fuliginosa* Lichens." *Journal of Food Science and Technology* 53, no. 6 (2016): 2804–16. https://doi.org/10.1007/s13197-016-2255-3.

[54] Legouin, Béatrice, Françoise Lohézic-Le Déváhat, Solenn Ferron, Isabelle Rouaud, Pierre Le Pogam, Laurence Cornevin, Michel Bertrand, and Joël Boustie. "Specialized Metabolites of the Lichen Vulpicida Pinastri Act as Photoprotective Agents." *Molecules* 22, no. 7 (December 2017): 1162. https://doi.org/10.3390/molecules22071162.

[55] Boustie, Joel, Sophie Tomasi, and Martin Grube. "Bioactive Lichen Metabolites: Alpine Habitats as an Untapped Source." *Phytochemistry Reviews* 10, no. 3 (2010): 287–307. https://doi.org/10.1007/s11101-010-9201-1.

[56] Rajan, Vinoshene, Saranyapiriya Gunasekaran, Surash Ramanathan, Vikneswaran Murugaiyah, Mohd. Samsudin, and Laily Din. "Biological Activities of Four Parmotrema Species of Malaysian Origin and Their Chemical Constituents." *Journal of Applied Pharmaceutical Science*, 2016, 036–43. https://doi.org/10.7324/japs.2016.60806.

[57] Kosanić, Marijana, Branislav Ranković, Tatjana Stanojković, Aleksandar Rančić, and Nedeljko Manojlović. "Cladonia Lichens and Their Major Metabolites as Possible Natural Antioxidant, Antimicrobial and Anticancer Agents." *LWT - Food Science and Technology* 59, no. 1 (2014): 518–25. https://doi.org/10.1016/j.lwt.2014.04.047.

[58] Kosanic Marijana, Branislav Rankovic, Tatjana Stanojkovic, Perica Vasiljevic, and Nedeljko Manojlovic. "Biological activities and chemical composition of lichens from Serbia" *EXCLI Journal* 13, no.13 (Nov 2014):1226-1238.

[59] Ranković, Branislav, Marijana Kosanić, Tatjana Stanojković, Perica Vasiljević, and Nedeljko Manojlović. "Biological Activities of Toninia Candida and Usnea Barbata Together with Their Norstictic Acid and Usnic Acid Constituents." *International Journal of Molecular Sciences* 13, no. 12 (December 2012): 14707–22. https://doi.org/10.3390/ijms131114707.

[60] Manojlović, Nedeljko, Branislav Ranković, Marijana Kosanić, Perica Vasiljević, and Tatjana Stanojković. "Chemical Composition of Three Parmelia Lichens and Antioxidant, Antimicrobial and Cytotoxic Activities of Some Their Major Metabolites." *Phytomedicine* 19, no. 13 (2012): 1166–72. https://doi.org/10.1016/j.phymed.2012.07.012.

[61] Cakmak, Kader Cetin and İlhami Gülçin. "Anticholinergic and antioxidant activities of usnic acid-an activity-structure insight"

Toxicology Reports 6, (November 2019): 1273–1280. https://doi.org/10.1016/j.toxrep.2019.11.003.

[62] Thadhani, Vinitha M., and Veranja Karunaratne. "Potential of Lichen Compounds as Antidiabetic Agents with Antioxidative Properties: A Review." *Oxidative Medicine and Cellular Longevity* 2017 (2017): 1–10. https://doi.org/10.1155/2017/2079697.

[63] Ou, B., D. Huang, M. Hampsch-Woodill, JA. Flanagan and EK. Deemer. "Analysis of antioxidant activities of common vegetables employing oxygen radical absorbance capacity (ORAC) and ferric reducing antioxidant power (FRAP) assays: a comparative study." *J Agric Food Chem* 50, no.11(2002):3122-8.

[64] Alzagameem, Abla, Basma Khaldi-Hansen, Dominik Büchner, Michael Larkins, Birgit Kamm, Steffen Witzleben, and Margit Schulze. "Lignocellulosic Biomass as Source for Lignin-Based Environmentally Benign Antioxidants." *Molecules* 23, no. 10 (2018): 2664. https://doi.org/10.3390/molecules23102664.

[65] Shrestha, Gajendra. *Exploring the Antibacterial, Antioxidant, and Anticancer Properties of Lichen Metabolites*, 2015.

[66] Pohanka, Miroslav, Hana Bandouchova, Jakub Sobotka, Jana Sedlackova, Ivana Soukupova, and Jiri Pikula. "Ferric Reducing Antioxidant Power and Square Wave Voltammetry for Assay of Low Molecular Weight Antioxidants in Blood Plasma: Performance and Comparison of Methods." *Sensors* 9, no. 11 (2009): 9094–9103. https://doi.org/10.3390/s91109094.

[67] Selvaraj, Gurudeeban, A Tinabaye and R Ananthi. "In Vitro Antioxidant Activities Of Salazinic Acid And Its Derivative Hexaacetyl Salazinic Acid." *International Journal of Research in Engineering and Technology* 04, no. 02 (2015): 345–55. https://doi.org/10.15623/ijret.2015.0402046.

[68] Verma, Neeraj, Bhaskar C Behera, Anjali Sonone, and Urmila Makhija. "Cell Aggregates Derived from Natural Lichen Thallus Fragments: Antioxidant Activities of Lichen Metabolites Developed in Vitro." *Natural Product Communications* 3, no. 11 (2008). 1911-1918 https://doi.org/10.1177/1934578x0800301124.

[69] Paz, G. Amo De, J. Raggio, M. P. Gómez-Serranillos, O. M. Palomino, E. González-Burgos, M. E. Carretero, and A. Crespo. "HPLC Isolation of Antioxidant Constituents from Xanthoparmelia Spp." *Journal of Pharmaceutical and Biomedical Analysis* 53, no. 2 (2010): 165–71. https://doi.org/10.1016/j.jpba.2010.04.013.

[70] Papadopoulou, Panagiota, Olga Tzakou, Constantinos Vagias, Panagiotis Kefalas, and Vassilios Roussis. "β-Orcinol Metabolites from the Lichen Hypotrachyna Revoluta." *Molecules* 12, no. 5 (December 2007): 997–1005. https://doi.org/10.3390/12050997.

[71] Gómez-Serranillos, M. Pilar, Carlos Fernández-Moriano, Elena González-Burgos, Pradeep Kumar Divakar, and Ana Crespo. "Parmeliaceae Family: Phytochemistry, Pharmacological Potential and Phylogenetic Features." *RSC Adv.* 4, no. 103 (December 2014): 59017–47. https://doi.org/10.1039/c4ra09104c.

[72] Singh, Shiv M., Purnima Singh, and Rasik Ravindra. "Screening of Antioxidant Potential of Arctic Lichens." *Polar Biology* 34, no. 11 (2011): 1775–82. https://doi.org/10.1007/s00300-011-1027-9.

[73] Lamuela-Raventós, Rosa M. "Folin-Ciocalteu Method for the Measurement of Total Phenolic Content and Antioxidant Capacity." *Measurement of Antioxidant Activity & Capacity*, 2017, 107–15. https://doi.org/10.1002/9781119135388.ch6.

[74] Blainski, Andressa, Gisely Lopes, and João De Mello. "Application and Analysis of the Folin Ciocalteu Method for the Determination of the Total Phenolic Content from Limonium Brasiliense L." *Molecules* 18, no. 6 (October 2013): 6852–65. https://doi.org/10.3390/molecules18066852.

[75] Rekatsina, Martina, Antonella Paladini, Alba Piroli, Panagiotis Zis, Joseph V. Pergolizzi, and Giustino Varrassi. "Pathophysiology and Therapeutic Perspectives of Oxidative Stress and Neurodegenerative Diseases: A Narrative Review." *Advances in Therapy* 37, no. 1 (2019): 113–39. https://doi.org/10.1007/s12325-019-01148-5.

[76] Cobley, James Nathan, Maria Luisa Fiorello, and Damian Miles Bailey. "13 Reasons Why the Brain Is Susceptible to Oxidative

Stress." *Redox Biology* 15, (2018): 490–503. https://doi.org/10.1016/j.redox.2018.01.008.

[77] Chen, Zhichun and Chunjiu Zhong. "Oxidative stress in Alzheimer's disease." *Neuroscience Bulletin* 30, no. 2 (April 2014): 271–81. https://doi.org/10.1007/s12264-013-1423-y271.

[78] Cheignon, C., M. Tomas, D. Bonnefont-Rousselot, P. Faller, C. Hureau and F. Collin. "Oxidative stress and the amyloid beta peptide in Alzheimer's disease." *Redox Biology* 14 (April 2018):450–54. https://doi.org/10.1016/j.redox.2017.10.014.

[79] Toulorge, Damien, Anthony H. V. Schapira, and Rodolphe Hajj. "Molecular Changes in the Postmortem Parkinsonian Brain." *Journal of Neurochemistry* 139 (May 2016): 27–58. https://doi.org/10.1111/jnc.13696.

[80] Dias, Vera, Eunsung Junn, and M. Maral Mouradian "The Role of Oxidative Stress in Parkinson's Disease." *Journal of Parkinson's Disease* 3, no.4 (2013): 461–91. https:/doi.org./10.3233/JPD-130230.

[81] Fernández-Moriano, Carlos, Pradeep Kumar Divakar, Ana Crespo, and M. Pilar Gómez-Serranillos. "In Vitro Neuroprotective Potential of Lichen Metabolite Fumarprotocetraric Acid via Intracellular Redox Modulation." *Toxicology and Applied Pharmacology* 316 (2017): 83–94. https://doi.org/10.1016/j.taap.2016.12.020.

[82] Toledo Marante, Francisco-Javier, A. García Castellano, F. Estévez Rosas, J. Quintana Aguiar and J. Bermejo Barrera. "Identification and Quantitation of Allelochemicals from the Lichen *Lethariella canariensis:* Phytotoxicity and Antioxidative Activity." *Journal of Chemical Ecology* 29, no. 9 (Sep 2003): 2049-71. https://doi.org/10.1023/a:1025682318001.

[83] Fernández-Moriano, Carlos, Pradeep Kumar Divakar, Ana Crespo, and M. Pilar Gómez-Serranillos. "Protective Effects of Lichen Metabolites Evernic and Usnic Acids against Redox Impairment-Mediated Cytotoxicity in Central Nervous System-like Cells." *Food and Chemical Toxicology* 105 (2017): 262–77. https://doi.org/10.1016/j.fct.2017.04.030.

[84] Fuente, M De La. "Effects of Antioxidants on Immune System Ageing." *European Journal of Clinical Nutrition* 56, no.3 (2002). https://doi.org/10.1038/sj.ejcn.1601476.

[85] Cernescu, Irina, Liliana Tarţău, Antonela Macavei and Catalina Elena Lupuşoru. "Experimental research on the effects of a Cetraria islandica extract on oxidative stress in laboratory animals" *Rev Med Chir Soc Med Nat Iasi,* 115, no. 3 (2011):899-904.

[86] Kotan, Elif, Lokman Alpsoy, Mustafa Anar, Ali Aslan, and Guleray Agar. "Protective Role of Methanol Extract of Cetraria Islandica (L.) against Oxidative Stress and Genotoxic Effects of AFB1 in Human Lymphocytes in Vitro." *Toxicology and Industrial Health* 27, no. 7 (2011): 599–605. https://doi.org/10.1177/07482337 10394234.

[87] Polat, Zühal, Elanur Aydın, Hasan Türkez, and Ali Aslan. "In Vitro Risk Assessment of Usnic Acid." *Toxicology and Industrial Health* 32, no. 3 (May 2013): 468–75. https://doi.org/10.1177/0748233713504811.

[88] Emsen, Bugrahan, Basak Togar, Hasan Turkez, and Ali Aslan. "Effects of Two Lichen Acids Isolated from Pseudevernia Furfuracea (L.) Zopf in Cultured Human Lymphocytes." *Zeitschrift Für Naturforschung* C 73, no. 7-8 (2018): 303–12. https://doi.org/10.1515/znc-2017-0209.

[89] Emsen, B, H Turkez, B Togar, and A Aslan. "Evaluation of Antioxidant and Cytotoxic Effects of Olivetoric and Physodic Acid in Cultured Human Amnion Fibroblasts." *Human & Experimental Toxicology* 36, no. 4 (2016): 376–85. https://doi.org/10.1177/0960327116650012.

[90] Pavlovic, Voja, Igor Stojanovic, Milka Jadranin, Vlatka Vajs, Iris Djordjević, Andrija Smelcerovic, and Gordana Stojanovic. "Effect of Four Lichen Acids Isolated from Hypogymnia Physodes on Viability of Rat Thymocytes." *Food and Chemical Toxicology* 51, (2013): 160–64. https://doi.org/10.1016/j.fct.2012.04.043.

[91] Ullah, Asmat, Abad Khan and Ismail Khan. "Diabetes mellitus and oxidative stress—A concise review." *Saudi Pharmaceutical Journal*

24, no. 5 (September 2016):547-53. https://doi.org/10.1016/j.jsps. 2015.03.013.

[92] Çolak, Suat, Fatime Geyikoglu, Hasan Türkez, Tülay Özhan Bakır, and Ali Aslan. "The Ameliorative Effect Of *Cetraria Islandica* against Diabetes-Induced Genetic and Oxidative Damage in Human Blood." *Pharmaceutical Biology* 51, no. 12 (2013): 1531–37. https://doi.org/10.3109/13880209.2013.801994.

[93] Çolak, Suat, Fatime Geyikoğlu, Tülay Özhan Bakır, Hasan Türkez, and Ali Aslan. "Evaluating the Toxic and Beneficial Effects of Lichen Extracts in Normal and Diabetic Rats." *Toxicology and Industrial Health* 32, no. 8 (2015): 1495–1504. https://doi.org/10.1177/0748233714566873.

[94] Çolak, Suat, Fatime Geyikoğlu, Ali Aslan, and Gülşah Yıldız Deniz. "Effects of Lichen Extracts on Haematological Parameters of Rats with Experimental Insulin-Dependent Diabetes Mellitus." *Toxicology and Industrial Health* 30, no. 10 (2012): 878–87. https://doi.org/10.1177/0748233712466130.

[95] Kattoor, AJ., NVK. Pothineni, D. Palagiri and JL Mehta. "Oxidative stress in Atherosclerosis." *Current Atherosclerosis Reports* 19, no. 11 (September 2017):42. https://doi.org/10.1007/s11883-017-0678-6.

[96] Sahin, E, S Dabagoglu Psav, I Avan, M Candan, V Sahinturk, and At Koparal. "Vulpinic Acid, a Lichen Metabolite, Emerges as a Potential Drug Candidate in the Therapy of Oxidative Stress–Related Diseases, Such as Atherosclerosis." *Human & Experimental Toxicology* 38, no. 6 (2019): 675–84. https://doi.org/10.1177/ 0960327119833745.

[97] Storz, Peter. "Oxidative Stress in Cancer." *Oxidative Stress and Redox Regulation*, 2013, 427–47. https://doi.org/10.1007/978-94-007-5787-5_15.

[98] Walton, Emma Louise. "The Dual Role of ROS, Antioxidants and Autophagy in Cancer." *Biomedical Journal* 39, no. 2 (2016): 89–92. https://doi.org/10.1016/j.bj.2016.05.001.

[99] Noda, Noriko and Hiro Wakasugi. "Cancer and Oxidative Stress." *Japan Medical Association Journal* 44, no. 12(2001): 535–39.

[100] Chen, Si, Zhuhong Zhang, Tao Qing, Zhen Ren, Dianke Yu, Letha Couch, Baitang Ning, et al. "Activation of the Nrf2 Signaling Pathway in Usnic Acid-Induced Toxicity in HepG2 Cells." *Archives of Toxicology* 91, no. 3 (January 2016): 1293–1307. https://doi.org/10.1007/s00204-016-1775-y.

[101] Sahu, Saura C., Margaret Amankwa-Sakyi, Michael W. Odonnell, and Robert L. Sprando. "Effects of Usnic Acid Exposure on Human Hepatoblastoma HepG2 Cells in Culture." *Journal of Applied Toxicology* 32, no. 9 (2011): 722–30. https://doi.org/10.1002/jat.1721.

[102] Su, Zu-Qing, Yu-Hong Liu, Hui-Zhen Guo, Chao-Yue Sun, Jian-Hui Xie, Yu-Cui Li, Jian-Nan Chen, Xiao-Ping Lai, Zi-Ren Su, and Hai-Ming Chen. "Effect-Enhancing and Toxicity-Reducing Activity of Usnic Acid in Ascitic Tumor-Bearing Mice Treated with Bleomycin." *International Immunopharmacology* 46 (2017): 146–55. https://doi.org/10.1016/j.intimp.2017.03.004.

[103] Prokopiev, Ilya, Galina Filippova, Eduard Filippov, Ivan Voronov, Igor Sleptsov, and Aliy Zhanataev. "Genotoxicity of (+) and (−) - Usnic Acid in Mice." *Mutation Research/Genetic Toxicology and Environmental Mutagenesis* 839 (2019): 36–39. https://doi.org/10.1016/j.mrgentox.2019.01.010.

[104] Goga, Michal, Martin Kello, Maria Vilkova, Klaudia Petrova, Martin Backor, Wolfram Adlassnig, and Ingeborg Lang. "Oxidative Stress Mediated by Gyrophoric Acid from the Lichen Umbilicaria Hirsuta Affected Apoptosis and Stress/Survival Pathways in HeLa Cells." *BMC Complementary and Alternative Medicine* 19, no. 1 (2019). https://doi.org/10.1186/s12906-019-2631-4.

[105] Emsen, Bugrahan, Ali Aslan, Basak Togar, and Hasan Turkez. "In Vitro antitumor Activities of the Lichen Compounds Olivetoric, Physodic and Psoromic Acid in Rat Neuron and Glioblastoma Cells." *Pharmaceutical Biology* 54, no. 9 (2015): 1748–62. https://doi.org/10.3109/13880209.2015.1126620.

[106] Santus, Pierachille, Angelo Corsico, Paolo Solidoro, Fulvio Braido, Fabiano Di Marco, and Nicola Scichilone. "Oxidative Stress and

Respiratory System: Pharmacological and Clinical Reappraisal of N-Acetylcysteine." *COPD: Journal of Chronic Obstructive Pulmonary Disease* 11, no. 6 (2014): 705–17. https://doi.org/10.3109/15412555.2014.898040.

[107] Su, Zu-Qing, Zhi-Zhun Mo, Jin-Bin Liao, Xue-Xuan Feng, Yong-Zhuo Liang, Xie Zhang, Yu-Hong Liu, et al. "Usnic Acid Protects LPS-Induced Acute Lung Injury in Mice through Attenuating Inflammatory Responses and Oxidative Stress." *International Immunopharmacology* 22, no. 2 (2014): 371–78. https://doi.org/10.1016/j.intimp.2014.06.043.

[108] Kim, So Youn, Chanseop Park, Hye-Jeong Jang, Bi-O Kim, Hee-Won Bae, In-Young Chung, Eun Sook Kim, and You-Hee Cho. "Antibacterial Strategies Inspired by the Oxidative Stress and Response Networks." *Journal of Microbiology* 57, no. 3 (2019): 203–12. https://doi.org/10.1007/s12275-019-8711-9.

[109] Peralta, Mariana, María Da Silva, María Ortega, José Cabrera, and María Paraje. "Usnic Acid Activity on Oxidative and Nitrosative Stress of Azole-Resistant Candida Albicans Biofilm." *Planta Medica* 83, no. 03/04 (2016): 326–33. https://doi.org/10.1055/s-0042-116442.

[110] Sinha, Sneha, Vivek Kumar Gupta, Parmanand Kumar, Rajiv Kumar, Robin Joshi, Anirban Pal, and Mahendra P. Darokar. "Usnic Acid Modifies MRSA Drug Resistance through down-Regulation of Proteins Involved in Peptidoglycan and Fatty Acid Biosynthesis." *FEBS Open Bio* 9, no. 12 (2019): 2025–40. https://doi.org/10.1002/2211-5463.12650.

In: A Comprehensive Guide ...
Editors: Silje A. Dahl et al.

ISBN: 978-1-53618-418-1
© 2020 Nova Science Publishers, Inc.

Chapter 3

PHYTOCHEMICALS: AN APPROACH TOWARDS ANTISICKLING ACTIVITY

Jaya Tiwari, Vijaylakshmi Jain and Pankaj Kishor Mishra[*]

Medical Biotechnology, Department of Biochemistry,
Pt Jawahar Lal Nehru Memorial Medical College, Raipur, India

ABSTRACT

Sickle cell disease (SCD) is the most prevalent inherited blood disorder affecting most parts of the world without any discrimination of ethics or racial standards. The patients undergo shortness of breath, heart palpitation, abdominal and muscle pain. Several managing SCD therapies have been proposed with treatment but all these treatments are ineffective or very expensive for the less fortunate population. Plant species used as folk medicines also display *in vitro* antisickling activity. Caricaceae (*Carica papaya*), Fabaceae (*Cajanus cajan* and *Crotalaria retusa*), Apocynaceae (*Raulwolfia vomitoria*, *Jatropha curcas*, *Euphorbia hirta*

---

[*] Corresponding Author's Email: pkjbiotech@gmail.com.

and *Picralima nitida*) and Euphorbiaceae (*Wrightia tinctoria* and *Alchornea cordifolia*) are prominently studied families of plants and their parts with their respective solvents for antisickling activity. Detection of various secondary metabolites like alkaloid, flavonoids, tannin, anthraquinone and many more bioactive compounds in these plants can be eminent cause for reducing sickle cells *in vitro* and can be further used for the development of therapeutic agent for cure of disease.

Keywords: antisickling, phytochemicals, alkaloid, flavonoids, tannin, anthraquinone

INTRODUCTION

The contribution of plants towards mankind with herbal remedies has been through several years (Tambe and Bhambar 2014). For many years medicine, nutrition, flavoring, fragrances, cosmetics, smoking, beverages, dyeing, repellents, and industrial purposes are accomplished by herbs. Since the pre-historic era, they have been the basis for nearly all medicinal therapy until synthetic drugs were developed (Djeridane et al. 2006). Overlooking as therapeutic agent medicinal plants have been a huge source of comprehension for a wide variety of chemical constituents which could be further developed as medication with specific preferences (Patil and Deshmukh 2016). Approximately 60-80% of total world population still fall back on traditional medicine for the ailment of common illness. Currently due to their versatile application plant derived substances are of great interest now days. Investigation of such plants to acknowledge their properties, safety and potency must be performed (Yadav and Agrawala 2011).

Ethnobotanical analysis promotes that plants are capable enough in treatment of various diseases and disorders including sickle cell disease (SCD). It is a genetic disorder resulting from inheritance of two abnormal allelomorphic genes that control the formation of β-chain of hemoglobin (Mpiana et al. 2010). Characterization of red blood cells with abnormal, rigid, sickle shape is autosomal recessive genetic disorder sickle cell anemia (Amujoyegbe et al. 2016). Substitution of glutamic acid with valine at the sixth position of beta-globin of hemoglobin chain results in SCD. The culmination of unusual hemoglobin is such that, under hypoxic condition, deoxy-Hbs molecule polymerize, forming rigid sickle cells which in turn causes the deformation of the normal disc biconcave RBC hence resulting in hemolytic anemia and blood vessel blockage prevailing the major causes of deaths in SCD (Avaligbe et al. 2012). Scarcity of oxygen, deoxyHbS protein polymerize, forming a rigid chain and induce the properties of SS-RBC shape. Usually, RBCs have intercellular Na^+ and Ca^{2+} low and K^+ and Mg^{2+} high and their pathways are regulated by cellular energy. Depletion of ATP and an increase in Ca^{2+} rise by three folds in SCD patients. Activation of Ca^{2+} dependent K^+ channel (Gardos channel) as cells contains few or no endocytic vesicle for storage of Ca^{2+}, dropping of Ca^{2+} accompanying the movement of Cl^- and water thereby resulting in dehydration and hemochrome formation. Several studies suggest that antisickling agent may act on inhibition of polymerization or inhibit RBC hemolysis (Vaishnava and Rangari 2019). Recurrences of pain with tiredness, pallor, respiratory infections, and blindness in old age are a common indication and shortness of breath, heart problems, and muscle tenderness also occurs with an increasing sickled erythrocyte (Dash et al. 2019). African negro population, the Mediterranean region, the middle east, and some regions of India are under the threat of this genetically inherited chronic disorder. The disease is quite prevalent in the area where malaria is endemic. Particularly, the first two years of life are highly fatal for children from this disease (Mgbemene and Ohiri 1999). The cost implementation, availability of necessary expertise, and finding suitable donors constitute major drawbacks in developing countries. Currently, in clinical practices drugs and chemical compounds are used in management but their side effects limit their clinical use (Singh et al. 2013). An alternative strategy in the management of SCD is now focused on the identification of novel antisickling agents mainly from medicinal plants. Indeed traditional medicine continues to play a very significant role in the medical primary health care implementation in developing countries (Ngbolua et al. 2015).

Thus this review aims to analyze natural products from frequently studied plants that belong to the family's apocynaceae, caricaceae, euphorbiaceae, and fabaceae for advances in anti-sickling agents.

THEORETICAL PERSPECTIVES

Demographics

High morbidity and mortality are being observed due to the presence of the HbS gene in at least 40 countries (Adam et al. 2019). Carriers of hemoglobinopathies have been estimated to be 5% of the world population and 7% of pregnant women. Throughout the world, 330,000 newborns are under the threat of major carriers of hemoglobinopathy and 275,000 of them have SCD (Ayung and Odame 2012). The emergence of disease in Sub-Saharan Africa, the Arabian Peninsula, and the Indian subcontinent and spreading of this population has uplifted the SCD frequency in regions where it was not earlier reported such as the USA, western and eastern Europe. It has been analyzed that more than 300,000 children are born every year with SCD about two-third of them are in Africa, Nigeria, India, and Democratic Republic Congo. In the USA affected individuals are close to 100,000 and 3000 newborns every year. It is a rapidly growing severe genetic disorder in the UK and Western Europe. A load of disease is getting worse because of insufficient health infrastructure, improper nutrition, and infectious diseases like malaria, tuberculosis, and HIV. In countries with ample resources and less than 1% disease load, over 90% of babies born with SCD continue to adulthood as newborn screening and the existence of all necessary care. In upcoming years or maybe till 2050 it has been anticipated that approximately 400,000 neonates will be born with SCD every year. Maharashtra, Gujarat, Tamil Nadu, Odisha, Punjab, Chhattisgarh are various prone areas of India with the prevalence of sickle gene in general among many tribal populations with 1-40% heterozygote variations (Colah et al. 2015; Thein and Thein 2016; Mburu and Odame 2019;Dash et al. 2019).

Clinical Manifestations

SCD patients endure unlike and variegated symptoms. Anemia, pain-related crisis, organ failure with crisis due to blocked blood vessels and damaged organs. Blood vessel occlusion which is triggered by membrane deformation results in a crisis. People apart from these also suffer from acute chest syndrome (ACS), acute splenic sequestration, hyperthermia, paripasm, vascular necrosis, proliferative retinopathy, aplastic crises, cholelithiasis, delayed growth, chronic pulmonary diseases, and chronic nephropathy. All these clinical manifestations of SCD do not appear until after the first sixth months of life, at which most of the HbF is replaced by HbS (Okpuzor et al. 2008).

Current Therapeutic Options in SCD

Blood Transfusion
Extensively red cell transfusion is used for treating SCD as more or less 50% of patients encounter transfusion at a different phase of their lives. Some common drawbacks of transfusion are that children with SCD are iron overload, alloimmunized, and transfusion-transmitted infection (Amrolia et al. 2003).

Hematopoietic Cell Transplantation (HCT) for SCD
Approaching 175 children with SCD has received matched sibling-donor HCT as a first allogeneic transplant for SCD was done in 1984 with proper pre-transplant preparations. The most limiting factor of this treatment is age and accessible suitable donors. Several hurdles after transplantation include intracerebral hemorrhage, graft-versus-host disease (GVHD), seizure, and gonadal dysfunction. Approximately after HCT, 91% of patients survive and 82% get healed from SCD (Hoppe and Walters 2001; Singh et al. 2013).

Gene Therapy

Human gene therapy for SCD consists of establishing safe and efficient retroviral packaging lines for gene transfer, retroviral vectors with the human β-globin gene, and a selectable marker. Achievements in the medication of human hemoglobin still require some more technical advances for increasing the efficiency of gene transfer and the level of gene expression (Bank et al. 1989).

Disease Modifier Agents

Developments in the system for diagnosis and care of SCD have elevated the life span of patients. Medicine which is being used for clinical management of SCD includes paracetamol, ibuprofen helps in reducing pain, penicillin are helpful in dealing with infections during the crisis, deferoxamine act as an iron chelator and folic acids are Hb inducer but although these medicines manifest various side effects like ulcer, liver function impairment, dizziness, abdominal cramps, diarrhea (Dash et al. 2019). Some of the usually directed medicines are listed in Table 1. Health protection of SCD substantially is based on preventive measures that are for eliminating painful crises by managing analgesics, antipyretics, oral antibiotics, and various epigenetic with alleviators of fetal Hb gene and many more techniques. The substitutive approach is required in conquering this disease through characterization and recognition of anti-sickling agents from medicinal plants because these techniques are either too costly for poor and not all worthwhile agents helps in dealing with the crises and may result with concomitant reactions or causing infection risk (Ngbolua et al. 2015; Igwe et al. 2017).

The desirability of researchers in finding antisickling agents from indigenous plants is a crucial point of study because of the restricted capability of drugs and techniques which are not affordable or currently unavailable in rural areas. Medicinal plants are a source of phytochemicals which might be useful in developing pharmaceutical drugs and can overcome the drawbacks (Nurain et al. 2017).

Therapeutic activity of phytochemicals in plant extracts are usually used in traditional practices. Substances or compounds which are found in providing relief are known as active principles which are dissimilar in every plant. Anthraquinones, flavonoids, glycosides, saponin, tannins, and many more are such examples of active principles. Plants also possess compounds like morphine, atropine, steroids, lactone, and volatile oils which can too have abilities in healing various diseases. These active compounds can be withdrawn by several methods such as infusion, syrups, concoction, decoction, infused oils, essential oils, ointments, and creams, therefore, making them necessary to investigated with its maximum importance (Sahu et al. 2012). Some of the frequently studied plants of families apocynaceae, caricaceae, euphorbiaceae, and fabaceae are listed in Tables 2, 3, 4, and 5 with their bioactive compounds.

Table 1. Disease modifiers with its uses and side effects

Drugs	Uses	Side affects	References
Hydroxyurea	Elevates HbF level, improves erythrocyte hydration, decrease red cell density	Myelosupression, hyperpigmentation, malignancy	(Halsey and Roberts 2003)
Clotrimazole	Blocking of Gardos channel, increase in intracellular K^+, reduces serum conjugate bilirubin concentration	Dysuria, nausea, diarrhea, chemical cystitis, and short-term increase in ALT/AST activity	(Bunn 1997; Brugnara 2003)
Sodium cromoglicate	Itstops Ca^{2+}-activated K^+Gardos channel and erythrocyte dehydration although its activity is unclear	Mild netropenia	(Toppet et al. 2000; Karimi et al. 2006)
Decitibine	Decreases hemolysis	Act as teratogen	(Saunthararajah et al. 2003; Molokie et al. 2017)
Senicapoc	Efficient in preventing RBC K^+ loss and dehydration in case of R352H mutation.	Painful episodes	(Rapetti-Mauss et al. 2016; Gee 2013)

Table 2. Phytochemical analysis of plants belonging to Apocynaceae

Plants	Plant parts	Solvents	Phytochemicals	References
Picralima nitida	Stem bark	Ethanol	Alkaloid, saponin, steroids, terpenoids, flavonoids	(Nkere and Iroegbu 2005)
	Seed (powder)		Alkaloid, tannin, glycosides, saponin, flavonoid, phenol	
	Pod (powder)		Alkaloid, flavonoids, saponin, cardiac glycosides, terpenoids	(Bruce et al. 2016)
	Leaf	Ethanol	Alkaloids, favonoids, tannins, saponin	(Ilodigwe et al. 2012)
	Root (powder)		Alkaloids, flavonoids, tannins, terpinoids, reducing sugars	(Ngaissona et al. 2016)
Rauwolfia vomitoria	Leaf	Methanol-water (50:50)	Alkaloids, tannins, flavonoids, saponin, cardiac glycosides	(Abere et al. 2014)
	Bark	Methanol	Alkaloids, tannins, saponin, terpenoids, flavonoids	(Ojo et al. 2012)
	Root	Ethanol	Alkaloids, flavonoids, tannin, saponin	(Oluwasina et al. 2017)
Wrightia tinctoria	Bark	Petroleum ether	Alkaloids, phenols, saponin, tannins	(Khyade and Vaikos 2011)
	Leaf	Ethyl acetate and methanol	Alkaloids, terpenoids, flavonoids, saponin	(Shankar et al. 2010)
	Root	Methanol	Steroid, terpenoids, tannins, saponin, coumarins, emodins	(Devi et al., 2014)
	Fruit (Peel+Cellulose)	Ethanolic	Tannins, phenols, steroid, anthraquinone	(Khandekar et al. 2013)

Table 3. Phytochemical analysis of plants belonging to Caricaceae

Plant	Plant part	Solvents	Phytochemicals	References
Carica papaya	Leaf	Water & Methanol (1:3)	Alkaloids, flavonoids, tannins, cardiac glycosides, anthraquinone, phlobatinins, saponin	(Imaga et al. 2010)
	Unripe fruit (powder)		Tannins, saponins, alkaloid, cardiac glycosides, anthraquinone	(Eke et al. 2014)
	Stem/bark	Methanol	Alkaloid, flavonoids, phenols, steroids, terprnoids, cardiac glycosides	(Nweri and Saidu 2013)
	Roots	Aqueous	Alkaloid, saponin, flavonoids	(Singh et al. 2018)

Table 4. Phytochemical analysis of plants belonging to family Euphorbiaceae

Plants	Plant parts	Solvents	Phytochemicals	References
Jatropha curcas	Leaf	Methanol	Alkaloid, saponin, tannin, terpenoid, steroid, glycosides	(Sharma et al. 2012)
	Root	Methanol	Alkaloid, saponin, tannin, steroid, glycosides, phenols	
	Stem	Methanol	Alkaloid, saponin, tannin, terpenoid, steroid, glycosides	
Euphorbia hirta	Leaves	Ethanol	Alkaloids, flavonoids, terpenoids, saponin	(Ahmad et al. 2017)
	Stem	Methanol	Terpenoid, alkaloid, reducing sugars	(Perumal et al. 2012)
	Flower	Methanol	Terpenoids, tannins, alkaloid, cardiac glycosides, reducing sugars	
Alchornea cordifolia	Leaf	Methanol	Alkaloids, saponin, tannins, flavonoids	(Osadebe et al. 2012)
	Stem	Ethanol (70:30 v/v in water)	Flavonoids, anthocynin, tannin, alkaloid, phenols	(Joseph et al. 2015)
	Root	Aqueous	Flavonoids, tannins, saponin, alkaloids	(Ishola et al. 2008)

Table 5. Phytochemical analysis of plants belonging to family Fabaceae

Plants	Plant part	Solvents	Phytochemicals	References
Crotolaria retusa	Seed	50% Hydroethanolic	General glycosides, saponin, tannins, alkaloids, flavonoids, sterols	(Anim et al. 2016)
	Pod		General glycosides, saponin, tannins, alkaloids, flavonoids, sterol, triterpenoids	
	Flower		General glycosides, saponin, tannins, alkaloids, flavonoids, sterols	
	Stem		General glycosides, saponin, tannins alkaloids, flavonoids, sterols	
	Leaf		General glycosides, saponin, tannins alkaloids, sterols	
Cajanus cajan	Leaf	Ethyl acetate	Steroids, cardiac glycosides, anthroquinone, saponin, alkaloid, phenols	(Mohanty et al. 2011)
	Seed	Petroleum Ether	Reducing sugars, alkaloids, terpenoids flavonoids, cardiac glycosides, anthraquinone, phenols	(Sahu et al. 2014)
	Stem	Petroleum ether	Tannins, saponin, alkaloids, terpenoids, flavonoids, cardiac glycosides, anthroquinone, phenols	
	Root	Methanol	Alkaloids, flavonoids, tannins, terpenoids, phenols, steroids, cardiac glycosides, anthraquinone, coumarins, saponins	(Devi et al. 2016)

RESULTS

Since past years ethno botanical assessment revealed that medicinal plants species belonging to different families can serve as a useful tool by traditional healers for treatment of SCD. Plant extracts which inhibit sickling of RBC or polymerization of deoxy form of HbS in hypoxic conditions and free radicals scavengers will alleviate all the chemical phenotype of SCD such as cell dehydration, hemolysis and damage (Mpiana et al. 2012). Secondary metabolites have primarily been known for their broad antimicrobial activity for the past 60 years. Various successful applications of secondary metabolites are antitumor agents, immune stimulants and also as cholesterol lowering agents. Unexpected functions are also observed in life threatening diseases such as prion disease, Alzheimer's, cancer, pulmonary diseases, cardiovascular diseases, parasitic and viral diseases (Vaishnav and Demain 2011). Phytochemicals like anthocyanin act as potent antioxidants and can reverse the pathological events leading to sickled form of RBC and under low oxygen tension may represent a rational explanation for managing SCD. Cells can be defended by haemolysis by terpenes and steroids. Anthraquinone are known to show antisickling activity *in-vitro*, they are also known to inhibit oxidative reaction occurring during pathophysiological changes in course of diseaseand also affect mean corpuscular volume (MCV). Saponin enhances natural resistance and recuperative capability of the body. Quinones are also assumed to change morphology of sickle RBC. Alkaloids are nerve stimulants, convulsions and muscle relaxants so it can also be useful in painful crisis in SCD. Coumarins are also known to act as antioxidants and also as antisickling agents. Flavonoids are also free radical scavengers (Kenner and Yves 1996; Mpiana et al. 2009a; Vaishnavaand Rangari 2019;Folashade andOmoregie 2013).

In apocynaceae family *Picralima nitida* seeds and leaves of *Rauwolfia vomitoria and Wrightia tinctoria* have shown antisickling activity (Abere et al. 2014;Osuala and Odoh 2017). *Carica papaya* leaves and unripe fruit of family caricaceae have been frequently studied for antisickling activity (Imaga et al. 2009, Mishra et al. 2018). Leaves of *Jatropha curcas* and *Alchornea cordifolia* as well as whole plant of *Euphorbia hirta* have been analysed for antisickling activity of family euphorbiaceae (Mpiana et al. 2009a; Mpiana et al. 2013). *Crotolaria retusa* leaves and *Cajanus cajan* seed and leaves have been also examined for antisickling activity (Mpiana et al. 2009b; Mishra et al. 2018). Presence of secondary metabolites as mentioned in Table 2, 3, 4 and 5 in plants may be helpful in reducing sickling or may be directly as well as indirectly involved in sickling and sickling related crisis. Compounds like ursolic acid, a well known triterpene, is also known to support antisickling activity from *O. gratissimum* leaves (Tshilanda et al. 2015). Butyl stearate an ester from *O. bsilicum* has also shown remarkable antisickling effect (Tshilanda et al. 2014). 5-hydroxymethyl-2-furfural (5HMF) a naturally occurring aromatic aldehyde from coffee, honey, dry fruits and flavouring agents by specifically binding to intercellular sickle haemoglobin without inhibition of plasma and tissue proteins and reduced sickling (Abdulmalik et al. 2005). Some literature study also supports that plant extracts having antisickling property may also be due to presence of some amino acids like phenylalanine, tyrosine, arginine, glutamic acid and asparagines as they are able to inhibit ratio of HbSS polymerization and Fe^{2+}/Fe^{3+} ratio (Singh et al. 2003).

CONCLUSION

In view of the fact that many decades ago the discovery of SCD and till now a readily available cure is not present although some outstanding achievements have been made in the area of gene therapy, transplantation technologies, some disease modifier agents and now stem cell investigations. This genetic disease affects most of the people from newborns to adult as they go through the crisis. People with low income and shortage of healthcare facilities also go through mental trauma also, so there is much need of cheap local herbal natural products that can help to improve the crises. The above study revealed the presence of certain bioactive compounds which may help in reversal of sickle erythrocyte and can be beneficial to mankind.

REFERENCES

Abdulmalik, O., Safo, M. K., Chen, Q., Yang, J., Brugnara, C., Ohene-Frempong, K. & Asakura, T. (2005). 5-hydroxymethyl-2-furfural modifies intracellular sickle haemoglobin and inhibits sickling of red blood cells. *British Journal of Haematology*, *128*(4), 552-561.

Abere, T. A., Ojogwu, O. K., Agoreyo, F. O. & Eze, G. I. (2014). Antisickling and toxicological evaluation of the leaves of *Rauwolfiavomitoria*Afzel (Apocynaceae). *Journal of Science and Practice of Pharmacy*, *1*(1), 11-15.

Adam, M. A., Adam, N. K. & Mohamed, B. A. (2019). Prevalence of sickle cell disease and sickle cell trait among children admitted to Al Fashir Teaching Hospital North Darfur State, Sudan. *BMC Research Notes*, *12*(1), 1-6.

Ahmad, W., Singh, S. & Kumar, S. (2017). Phytochemical screening and antimicrobial study of *Euphorbia hirta* extracts. *Journal of Medicinal Plants Studies*, *5*, 183-186.

Amrolia, P. J., Almeida, A., Halsey, C., Roberts, I. A. & Davies, S. C. (2003). Therapeutic challenges in childhood sickle cell disease Part 1: current and future treatment options. *British Journal of Haematology*, *120*(5), 725-736.

Amujoyegbe, O. O., Idu, M., Agbedahunsi, J. M. & Erhabor, J. O. (2016). Ethnomedicinal survey of medicinal plants used in the management of sickle cell disorder in Southern Nigeria. *Journal of Ethnopharmacology*, *185*, 347-360.

Anim, M. T., Larbie, C., Appiah-Opong, R., Tuffour, I.,Owusu, K. B. & Aning, A. (2016). Phytochemical, antioxidant and cytotoxicity ofhydroethanolic extracts of *Crotalaria retusa*L. *World Journal of Pharmaceutical Research*, *5*, 162-179.

Avaligbe, C. T., Gbenou, J. D., Kpoviessi, D. S., Gbaguidi, F., Moudachirou, M., Accrombessi, G. C. & Gbeassor, M. (2012). Assessment of anti-sickling properties of extracts of plants used in the traditional treatment of sickle cell disease in Benin. *European Journal of Scientific Research*, *87*, 100-108.

Ayung, B. & Odame, I. (2012). A global perspective on sickle cell disease. *Pediatric Blood and Cancer*, *59*(2), 386-390.

Bank, A., Markowitz, D. & Lerner, N. (1989). Gene transfer. A potential approach to gene therapy for sickle cell disease. *Annals of the New York Academy of Sciences*, *565*, 37-43.

Bruce, S. O., Onyegbule, F. A. & Ihekwereme, C. P. (2016). Evaluation of the hepato-protective and anti-bacterial activities of ethanol extract of *Picralimanitida* seed and pod. *Journal of Phytomedicine and Therapeutics*, *15*(2), 1-22.

Brugnara, C. (2003). Sickle cell disease: from membrane pathophysiology to novel therapies for prevention of erythrocyte dehydration. *Journal of Pediatric Hematology/Oncology*, *25*(12), 927-933.

Bunn, H. F. (1997). Pathogenesis and treatment of sickle cell disease. *New England Journal of Medicine*, *337*(11), 762-769.

Colah, R. B., Mukherjee, M. B., Martin, S. & Ghosh, K. (2015). Sickle cell disease in tribal populations in India. *The Indian Journal of Medical Research*, *141*(5), 509.

Dash, D. K., Sahu, A., Panik, R., Mishra, K., Kashyap, P. & Tripathi, V. (2019). Comparative screening of anti-sickling reported and traditional herbs found in Chhattisgarh. *International Journal of Pharmaceutical Chemistry and Analysis*, *6*, 88-94.

Devi, P. S., Satyanarayana, B. & Naidu, M. T. (2014). Phytochemical screening for secondary metabolites in *Boswelliaserrata* Roxb. And *Wrightiatinctoria* (Roxb.) R. Br. *Notulae Scientia Biologicae*, 6(4), 474-477.

Devi, R. R., Premalatha, R. & Saranya, A. (2016). Comparative analysis of phytochemical constituents and antibacterial activity of leaf, seed and root extract of *Cajanuscajan* (L.) Mill sp. *International Journal of Current Microbiology and Applied Sciences*, *5*(3), 485-494.

Djeridane, A., Yousfi, M., Nadjemi, B., Boutassouna, D., Stocker, P. & Vidal, N. (2006). Antioxidant activity of some Algerian medicinal plants extracts containing phenolic compounds. *Food Chemistry*, *97*(4), 654-660.

Eke, O. N., Augustine, A. U. & Ibrahim, H. F. (2014). Qualitative analysis of phytochemicals and antibacterial screening of extracts of *Carica papaya* fruits and seeds. *International Journal of Modern Chemistry*, 6, 48-45.

Folashade, K. O. & Omoregie, E. H. (2013). Chemical constituents and biological activity of medicinal plants used for the management of sickle cell disease-A review. *Journal of Medicinal Plants Research*, 7(48), 3452-3476.

Gee, B. E. (2013). Biologic complexity in sickle cell disease: implications for developing targeted therapeutics. *The Scientific World Journal*, *694146*.

Halsey, C. & Roberts, I. A. (2003). The role of hydroxyurea in sickle cell disease. *British Journal of Haematology*, 120(2), 177-186.

Hoppe, C. C. & Walters, M. C. (2001). Bone marrow transplantation in sickle cell anemia. *Current Opinion in Oncology*, 13(2), 85-90.

Igwe, C. U., Alisi, C. S., Nwaoguikpe, R. N., Ojiako, A. O. & Onwuliri, V. A. (2017). Relationship between antisickling potency of edible African plants and their amino acid compositions. *Nigerian Journal of Biochemistry and Molecular Biology*, 32(2), 110-119.

Ilodigwe, E. E., Okoye, G. O., Mbagwu, I. S., Agbata, C. A. & Ajaghaku, D. L. (2012). Safety evaluation of ethanol leaf extract of *Picralimanitida* Stapf (Apocynaceae). *International Journal of Pharmacology and Therapeutics*, 2(4), 6-17.

Imaga, N. A., Gbenle, G. O., Okochi, V. I., Adenekan, S., Duro-Emmanuel, T., Oyeniyi, B. & Ekeh, F. C. (2010). Phytochemical and antioxidant nutrient constituents of *Carica papaya* and *Parquetinanigrescens* extracts. *Scientific Research and Essays*, 5, 2201-2205.

Imaga, N. A., Gbenle, G. O., Okochi, V. I., Akanbi, S. O., Edeoghon, S. O., Oigbochie, V. &Bamiro, S. B. (2009). Antisickling property of *Carica papaya* leaf extract. *African Journal of Biochemistry Research*, 3, 102-106.

Ishola, I. O., Ashorobi, R. B. & Adoluwa, O. (2008). Evaluation of antistress potential and phytochemical constituents of aqueous root extract of *Alchorneacordifolia*. *Asian Journal of Scientific Research*, 1(4), 476-80.

Joseph, N., Sorel, N. E. M., Kasali, F. M. & Emmanuel, M. M. (2015). Phytochemical screening and antibacterial properties from extract of *Alchorneacordifolia* (Schumach. & Thonn.) Müll. Arg. *Journal of Pharmacognosy and Phytochemistry*, *4*(3), 176-180.

Karimi, M., Zekavat, O. R., Sharifzadeh, S. & Mosavizadeh, K. (2006). Clinical response of patients with sickle cell anemia to cromolyn sodium nasal spray. *American Journal of Hematology*, *81*(11), 809-816.

Kenner, D. L. & Yves, R. M. D. (1996). Botanical Medicine. *A European Professional Perspective.*,*147*, 487.

Khandekar, U. S., Ghongade, R. A. & Mankar, M. S. (2013). Phytochemical investigation on *Wrightiatinctoria* fruit. *International Journal of Pharmacognosy and Phytochemical Research*, *5*, 41-44.

Khyade, M. S. & Vaikos, N. P. (2011). Comparative phytochemical and antibacterial studies on the bark of *Wrightiatinctoria* and *Wrightiaarborea*. *International Journal of Pharma and Bio Sciences*, *2*(1), 176-181.

Mburu, J. & Odame, I. (2019). Sickle cell disease: Reducing the global disease burden. *International Journal of Laboratory Hematology*, *41*, 82-88.

Mgbemene, C. N. & Ohiri, F. C. (1999). Anti-sickling potential of *Terminalia catappa* leaf extract. *Pharmaceutical Biology*, *37*(2), 152-154.

Mishra, P. K., Sharma, S., Jain, V., Tiwari, J., Mishra, M., Patra, P. K. & Khodiar, P. K. (2018). Antisickling and antioxidant relevance of twelve ethnomedicinal plants. *Medicinal Plants-International Journal of Phytomedicines and Related Industries*, *10*(3), 226-235.

Mohanty, P. K., Chourasia, N., Bhatt, N. K. & Jaliwala, Y. A. (2011). Preliminary phytochemical screening of *Cajanuscajan* Linn. *Asian Journal of Pharmacy and Technology*, *1*(2), 49-52.

Molokie, R., Lavelle, D., Gowhari, M., Pacini, M., Krauz, L., Hassan, J., Ibanez, V., Ruiz, M. A., Ng, K. P., Woost, P., Radivoyevitch, T., Pacelli, D., Fada, S., Rump, M., Hsieh, M., Tisdale, J. F., Jacobberger, J., Phelps, M., Engel, J. D., Saraf, S. & Saunthararajah, Y. (2017). Oral tetrahydrouridine and decitabine for non-cytotoxic epigenetic gene regulation in sickle cell disease: A randomized phase 1 study. *PLoS Medicine*, *14*(9), e1002382.

Mpiana, P. T., Lombe, B. K., Ombeni, A. M., Tshibangu, D. S., Wimba, L. K., Tshilanda, D. D. & Muyisa, S. K. (2013). *In vitro* sickling inhibitory effects and anti-sickle erythrocytes hemolysis of *Diclipteracolorata* CB Clarke, *Euphorbia hirta* L. and *Sorghum bicolor* (L.) Moench. *Open Journal of Blood Diseases*, *3*, 43-48.

Mpiana, P. T., Mudogo, V., Tshibangu, D. S. T., Ngbolua, K. N., Tshilanda, D. D. & Atibu, E. K. (2009a). Antisickling activity of anthocyanins of *Jatropha curcas* L. *Recent Progress in Medicinal Plants*, *25*, 101-108.

Mpiana P. T., Mudogo, V., Ngbolua, K. N., Tshibangu, D. S. T., Atibu, E. K., Kitwa, E. K. & Kanangila, A. B. (2009b). *In vitro* antisickling activity of anthocyanins extracts of *Vignaunguiculata* (L.) Walp. *Chemistry and medicinal value. Houston: Studium Press LLC*, 75-82.

Mpiana, P. T., Ngbolua, K. N., Mudogo, V., Tshibangu, D. S. T., Atibu, E. K., Mbala, B. M. & Makelele, L. T. (2012). The potential effectiveness of medicinal plants used for the treatment of sickle cell disease in the Democratic Republic of Congo folk medicine: A review. *Progress in Traditional and Folk Herbal Medicine*, *1*, 1-11.

Mpiana, P. T., Makele, L. K., Oteko, R. W., Akota, B. B., Tshbangu, M. T., Ngbolua, D. S. T., Mbala, B. M., Atibu E. K. & Nshimba, S. M. (2010).Antisickling activity of medicinal plants used in management of sickle cell disorder in Toshopo district, DR Congo. *Australian Journal of Medical Herbalism*, *22*, 132-137.

Ngaissona, P., Namkona, F. A., Koane, J. N., Tsiba, G., Syssa-Magale, J. L. & Ouamba, J. M. (2016). Phytochemical screening and evaluation of the antioxidant activity of the polar extracts *Picralimanitida* Stapf. (Apocynaceae) family. *Journal of Pharmacognosy and Phytochemistry*, *5*(4), 198.

Ngbolua, K. N., Tshibangu, D. S. T., Mpiana, P. T., Mihigo, S. O., Mavakala, B. K., Ashande, M. C. & Muanyishay, L. C. (2015). Anti-sickling and antibacterial activities of some extracts from *Gardenia ternifolia* subsp. jovis-tonantis (Welw.) Verdc. (Rubiaceae) and

Uapacaheudelotii Baill. (Phyllanthaceae). *Journal of Advances in Medical and Pharmaceutical Sciences*, 10-19.

Nkere, C. K. & Iroegbu, C. U. (2005). Antibacterial screening of the root, seed and stem bark extracts of *Picralimanitida*. *African Journal of Biotechnology*, 4(6), 522-526.

Nurain, I. O., Bewaji, C. O., Johnson, J. S., Davenport, R. D. & Zhang, Y. (2017). Potential of three ethnomedicinal plants as antisickling agents. *Molecular Pharmaceutics*, 14(1), 172-182.

Nweri, C. G. & Saidu, A. N. (2013). Phytochemical Screening and effects of methanol extract of *Carica papaya* stem bark in alloxan induced Diabetic Rats. *Journal of Emerging Trends in Engineering and Applied Sciences*, 4(6), 819-822.

Ojo, O. O., Ajayi, S. S. & Owolabi, L. O. (2012). Phytochemical screening, anti-nutrient composition, proximate analyses and the antimicrobial activities of the aqueous and organic extracts of bark of *Rauvolfiavomitoria* and leaves of *Peperomiapellucida*. *International Research Journal of Biochemistry and Bioinformatics*, 2(6), 127-134.

Okpuzor, J., Adebesin, O., Ogbunugafor, H. & Amadi, I. (2008). The potential of medicinal plants in sickle cell disease control: a review. *International Journal of Biomedical and Healthcare Science*, 4(2), 47-55.

Oluwasina, O. O., Olagboye, S. A., Olaiya, A. & Hassan, F. G. (2017). Comparative study on phytochemical quantification and antimicrobial activity of *Raufolvia Vomitoria* laves, seeds and root extracts. *FUTA Journal of Research in Sciences*, 13, 10-16.

Osadebe, P. O., Okoye, F. B., Uzor, P. F., Nnamani, N. R., Adiele, I. E. & Obiano, N. C. (2012). Phytochemical analysis, hepatoprotective and antioxidant activity of *Alchorneacordifolia* methanol leaf extract on carbon tetrachloride-induced hepatic damage in rats. *Asian Pacific Journal of tropical medicine*, 5(4), 289-293.

Osuala, F. N. & Odoh, U. E. (2017). Isolation and characterization of anti-sickling bioactive compounds from seeds of *Picralima Nitida* Stapf (Apocynaceae). *Asian Journal of Pharmaceutical Technology and Innovation*, 5(23), 79-85.

Patil, U.S. & Deshmukh, O.S. (2016). Preliminary phytochemical screening of six medicinal plants used in traditional medicine. *International Journal of Pharma and Bio Sciences*, 7, 77-81.

Perumal, S., Pillai, S., Cai, L. W., Mahmud, R. & Ramanathan, S. (2012). Determination of minimum inhibitory concentration of *Euphorbia hirta* (L.) extracts by tetrazolium microplate assay. *Journal of Natural Products*, 5(2), 68-76.

Rapetti-Mauss, R., Soriani, O., Vinti, H., Badens, C. & Guizouarn, H. (2016). Senicapoc: a potent candidate for the treatment of a subset of hereditary xerocytosis caused by mutations in the Gardos channel. *Haematologica*, 101(11), e431.

Sahu, M., Singh, V., Yadav, S. & Harris, K. K. (2012). Plant extracts with antisickling propensities: a feasible succour towards sickle cell disease management-a mini review. *Journal of Phytology.*, 4, 24-29.

Sahu, M., Vermaand, D. & Harris, K. K. (2014). Phytochemical analysis of the leaf, stem and seed extracts of *Cajanuscajan* L (Dicotyledoneae: Fabaceae). *World Journal of Pharmacy and pharmaceutical sciences*, 3(8), 694-733.

Saunthararajah, Y., Hillery, C. A., Lavelle, D., Molokie, R., Dorn, L., Bressler, L. & DeSimone, J. (2003). Effects of 5-aza-2′-deoxycytidine on fetal hemoglobin levels, red cell adhesion, and hematopoietic differentiation in patients with sickle cell disease. *Blood*, 102(12), 3865-3870.

Shankar, S. R., Rangarajan, R., Sarada, D. V. L. & Kumar, C. S. (2010). Evaluation of antibacterial activity and phytochemical screening of *Wrightiatinctoria* L. *Pharmacognosy Journal*, 2(14), 19-22.

Sharma, A. K., Gangwar, M., Tilak, R., Nath, G., Sinha, A. S. K., Tripathi, Y. B. & Kumar, D. (2012). Comparative *in vitro* antimicrobial and phytochemical evaluation of methanolic extract of root, stem and leaf of *Jatropha curcas* Linn. *Pharmacognosy Journal*, 4(30), 34-40.

Singh, B., Bhat, T. K. & Singh, B. (2003). Potential therapeutic applications of some antinutritional plant secondary metabolites. *Journal of Agricultural and Food Chemistry*, 51(19), 5579-5597.

Singh, M., Kaur, M., Dangi, C. B. S. & Singh, H. (2013). *In vitro* antidrepanocytary activity (anti-sickle cell anemia) of phytochemicals: natural source of medicine for sickle cell disease. *International Journal of Medicinal Plants*, *105*, 174-203.

Singh, P., Tanwar, N., Saha, T., Gupta, A. & Verma, S. (2018). Phytochemical screening and analysis of *Carica papaya*, *Agave americana* and *Piper nigrum*. *International Journal of Current Microbiology and Applied Sciences*, *7*(2), 1786-94.

Tambe, V. D. & Bhambar, R. S. (2014). Estimation of total phenol, tannin, alkaloid and flavonoid in *Hibiscus tiliaceus* Linn. wood extracts. *Journal of Pharmacognosy and Phytochemistry*, *2*(4), 41-47.

Thein, M. S. & Thein, S. L. (2016). World Sickle Cell Day 2016: A time for appraisal. *The Indian Journal of Medical Research*, *143*(6), 678.

Toppet, M., Fall, A. B., Ferster, A., Fondu, P., Melot, C., Vanhaelen-Fastre, R. & Vanhaelen, M. (2000). Antisickling activity of sodium cromoglicate in sickle-cell disease. *The Lancet*, *356*(9226), 309.

Tshilanda, D. D., Mpiana, P. T., Onyamboko, D. N. V., Mbala, B. M., Tshibangu, D. S. T., Bokolo, M. K. & Kasonga, T. K. (2014). Antisickling activity of butyl stearate isolated from *Ocimumbasilicum* (Lamiaceae). *Asian Pacific Journal of Tropical Biomedicine*, *4*(5), 393-398.

Tshilanda, D. D., Onyamboko, D. N., Babady-Bila, P., Tshibangu, D. S. & Mpiana, P. T. (2015). Anti-sickling activity of ursolic acid isolated from the leaves of *Ocimumgratissimum*L.(Lamiaceae). *Natural Products and Bioprospecting*, *5*(4), 215-221.

Vaishnav, P. & Demain, A. L. (2011). Unexpected applications of secondary metabolites. *Biotechnology Advances*, *29*(2), 223-229.

Vaishnava, S. & Rangari, V. D. (2019). *In-vitro* anti-sickling activity of selected medicinal plant to explore herbal remedies for sickle cell anemia. *International Journal of Pharma Research and Health Sciences*, *7*, 2909-2914.

Yadav, R. N. S. & Agarwala, M. (2011). Phytochemical analysis of some medicinal plants. *Journal of Phytology*, *3*, 10-14.

Chapter 4

PHYTOCHEMICAL POTENCY AND BIOCOMPATIBLE COMPARISON OF GREEN AND ORGANIC SOLVENT

Vijaylakshmi Jain, Jaya Tiwari and Pankaj Kishor Mishra[*]

Medical Biotechnology, Department of Biochemistry,
Pt Jawahar Lal Nehru Memorial Medical College, Raipur, India

Abstract

Medicinal plants are the repository of therapeutical drugs in terms of secondary metabolite or phytochemicals. The extraction of this secondary metabolite is one of the major steps of the whole process. Solvents despite being part of the whole process are not part of the composition and formulation process. Protracted exposure to solvents has adverse effects on all living organisms, causing detrimental effects on the body. Reduced use or replacement of such solvent with less toxicity and the harmful effect is the need of today. Green technology or green solvent systems has brought a revolutionary change in the toxin world. A less toxic, easy accessibility,

---

[*] Corresponding Author's Email: pkjbiotech@gmail.com.

higher possibility of reuse and a great efficient approach is how green solvent is characterized. Solvent efficiency is highly impacted by the type of solvent and so the phytochemical constituent. This chapter describes the efficiency and comparative analysis of green and conventional solvent systems concerning the three classes of phytochemicals i.e., phenols, alkaloids, and flavonoids. It also accounts for the antimicrobial activity against gram-negative, gram-positive and fungal strains of both the solvent system and its efficiency. The implications of the findings for phytochemicals show that extraction efficiency of phytochemicals was found high in the green solvent than organic ones and antimicrobial activity was also observed high in organic than the conventional one. The study discusses the solvent, solvent system, phytochemicals, antimicrobial infections and future perspective to the whole outcome.

Keywords: phytochemicals, microbial infection, antimicrobial activity, green solvent, efficiency

INTRODUCTION

Plants are referred to as the originator of natural ingredients and source of secondary metabolites such as tannin, alkaloids, flavonoids, phenol, glycosides, volatile oil, etc. composed from the different plant parts [1]. Medicinal plants are like a reservoir of therapeutic phytochemicals or secondary metabolites, which can further lead to novel isolations and applications in the form of nutraceuticals, fragrances, biopharmaceuticals, biofuels, and other products. There is an escalating scientific interest in the extraction and study of secondary metabolites in plants like biosynthetic, biochemical, chemotaxonomic, ecological, phytochemical, pharmacological, medicinal, traditional, and plant tissue culture studies. Phytochemicals are bioactive compounds found in plants, which work with nutrients and dietary fiber to protect against diseases. Most of them have antioxidant activity, which reduces the risk of many diseases.

Secondary metabolites have some curative activity against several ailments in humans and therefore, could explain the traditional use of the medicinal plant for the treatment of some basic illnesses [2, 3]. According to the World Health Organization (WHO), medicinal plants would be the best source to get a variety of drugs. Not only in ayurveda but also in modern medicine, the plant leaves are used for several bioactive ingredients. About 80% of people from developed countries use traditional medicines, which has compounds derived from medicinal plants. However, such plants should be investigated to better understand their property, safety, and efficiency [4].

The medicinal value of plants can be assessed by their phytochemical constituents and secondary metabolites content like alkaloids, flavonoids, phenolic, tannins, glycosides, volatile oils, terpenoids, etc. [5]. The sample preparation for any analysis plays an indispensable role when the isolation of constituent trace and ultra-trace compounds are concerned. Sample handling is the most basic and crucial part of any analysis which is aimed to develop efficient, faster, safer, and more eco-friendly methods. The study of natural products from plants begins with the pre-extraction and extraction procedures. Correspondingly methods of extracting phytochemicals from its onsets are the leading focus of the investigation. Current and informed isolation and chemical purification procedures of solvent extraction, utilize solvent polarity as a major separation technique. These methods frequently include the use of ethyl acetate, phenol, acetone, chloroform, aqueous, and other accessions [6, 7].

Microbial infections over the last 25 years have risen to a dark age in infectious history, where microbes are developing more rapidly and powerfully by developing resistance to antibiotics, and major new antibiotics are not being manufactured by pharma industries to deal with the current situations. Infectious diseases dominate the World Health Organization (WHO), 2019 list of top 10 threats to global health [8]. Therefore, a sudden urge to prevent infections from further spreading is needed. Bioactive compounds present in medicinal plants have been studied for their therapeutic purpose and can act like the ones.

This chapter intends to provide guidance on the basic phytochemistry and to study the isolated secondary metabolite constituents in both green and conventional solvent extraction from plants and the efficiency of the solvent extraction to be able to produce the most effective treatments and drugs by studying its antimicrobial susceptibility.

GREEN VS. ORGANIC SOLVENTS

Solvents

Solvents extensively defines the environmental performance of whole processes in chemical and other related industries, as they are not only present in the formulation and purification process but also pose in the final product, thus are the part of the whole cycle of production having its impact on cost, safety and health issues. Solvents are substances capable of dissolving another substance and forming a dispersed mixture or solution. They should be considered and chosen carefully beforehand like, ability to dissolve the secondary metabolite under study, easily removable, non-toxic, inert, and less volatile. Solvents can be categorized broadly in two groups:

1. Polar- Hydrophilic and soluble in water
2. Non-polar- Lipophilic and soluble in organic media [9].

Organic Solvents

Extraction solvents affect the separation capability of bioactive compounds from plant material and other sources. Organic solvents are generally classified as aliphatic hydrocarbons, aromatic hydrocarbon, cyclic hydrocarbon, halogenated hydrocarbons, amines, ketones, esters, ether, aldehydes, alcohols etc. It is usually possible to select a solvent by not acquiring a large amount of physical work as the properties of commonly used solvents are available [10].

Despite being the part of the whole process, solvents are neither in the composition system nor in the active part of formulation process, hereof the use of volatile, flammable, toxic and eco-unfriendly solvents, where it doesn't aid characteristically in any form is futile, yet, this unyielding consequence ends up to the beneficial facet of the solvent (Table 1).

Several attempts have been made to replace the undesirable solvents with related compounds to reduce the after effects of the solvent. Tremendous dose of many organic solvents leads to acute reversible narcotic state and can induce chronic, long lasting changes in nervous system function. Thus, Benzene was replaced by toluene as the former is well recognized as carcinogen to humans, affecting the bone marrow that causes leukopenia and thrombocytopenia [19, 20]. Similarly, carbon tetrachloride (CCl_4), one of the greenhouse gas and ozone depletory was regulated for restricted use [21] and was replaced with the halogenated solvents like chloroform and dichloromethane (DCM), and various polycyclic aromatic hydrocarbons persist in environment and has proved to cause carcinogenic and mutagenic effect and are also potent immunosuppressant [22].

Although with such measures, the decision were proved to be short-sighted as prolonged usage of toluene cause organ damage, damage to womb, chronic inhalation of pure toluene used in paints, thinner, glue and as cleansing agent often leads towards cerebellar dysfunction [22]. Chloroform and DCM were also listed as carcinogens and are both hepatotoxic and nephrotoxic in most animal species [23, 24] in WHO list. Moreover, DCM one of the short-lived halogen was proved as ozone depletory [25].

Table 1. Characteristics properties of organic solvents

Characteristics	Beneficial Aspect	Risks	References
Volatility	Permits recovery and purification of solvent by distillation	Unwanted air exposure, working individual exposure and health issues	11
Amide nature	High polarity- Dissolves varied substrates, accelerates reaction	Reproductive toxicity	12,13
Hydrocarbon nature	Dissolve oil during extraction and helps in separation	Highly combustible, low water solubility linked to bioaccumulation and aquatic toxicity	14-18

Industrially and commercially important solvents like n-hexane and methyl ketone undergo neurotoxic conversion in experimental animals [26]. Ethyl acetate, a most common ester occurs as a volatile liquid at room temperature is used for flavoring, confectionery, perfume manufacturing as well as in aroma therapy [27]. Alcohols as a solvent are used in organic synthesis, cleansing agent and as a reagent among which methanol, ethylene glycol and isopropyl alcohol are very common and expresses toxic effect [28].

High levels of pollutant due to discharge and accumulation of chemicals has lead to lower the number of wildlife, damage the ecosystem and are possessing threat to human health. The mentioned facts and literatures on organic solvent have driven attention among academia and industries for use of green solvent. Green chemistry aims less use of harmful solvent or replacing them with slighter toxic ones [29].

One of the studies forecasted the increased demands of solvents in upcoming years [30]. Although usage of strong compounds will decrease, the heightened industrial needs will lead to increased demands. With very poor toxicological and environmental profiles, volatile, neurotoxic, nephrotoxic, carcinogenic nature, and harmful risks on human life, there is a need for an alternative solution which is not only economically favorable but with broad industrial implementations. Around twenty million tons of leftovers from organic solvents are discharged in the environment and polluting it every year [31]. The prolonged exposure to solvent can damage the respiratory and nervous system and all the other systems present in a living organism.

Green Solvent

A green solvent can be defined as an idea aiming for replacement of commonly used solvent in any chemical process or eco-friendly solution with no or minimizations of ecological footprints, toxicity, risks to health and safety. Solvent efficiency is dependent highly on the solvent type. They are mainly characterized on the basis of low toxicity, easy accessibility, great efficiency and reusability [32]. Green solvent improves chemical process, less solvent usage and decreased processing steps [33].

To achieve the sustainable eco-friendly development of chemical industry, Paul Anastas and John Warner in 1998 put forward the twelve principles of Green chemistry to provide a greener chemical process or product [34]. Followings are the 12 principles:

1. *Prevention:* It is better to prevent waste than to treat or clean up waste after it has been created.
2. *Atom Economy:* Synthetic methods should be designed to maximize the infusion of all materials used in the process into the final product.
3. *Less Hazardous Chemicals:* Synthetic methods should be designed to process and generate substances with little or no toxicity to human health and environment wherever possible.
4. *Designing Safer Chemicals:* Chemicals employed during the process should be of highest efficacy and lowest toxicity.
5. *Safer Solvents and Auxiliaries:* Auxiliary substances like solvents, separation agents, etc. which uses less energy, least toxicity, least environment impacts and no major safety impacts should be used.
6. *Design for Energy Efficiency:* The chemical process should be carried out at ambient temperature and pressure. Energy required as per their environmental and economic impacts should be consumed with minimization as its basic principle.
7. *Use of Renewable Feedstock's:* Renewable raw materials or feedstock should be used.
8. *Reduce Derivatives:* Unnecessary derivatization of chemical process like blocking groups, protection/deprotection, temporary modifications, etc) should be minimized or avoided if possible which can create more wastes.
9. *Catalysis and new Catalytic Reagents:* Catalyst should be used to increase the economic efficacy of the chemical process which can be recycled indefinitely.
10. *Design Products for Degradation*: Chemical products should be designed to be harmless and biodegradable in the end and do not persist in the environment.
11. *Real-time analysis for Pollution Prevention*: The effective application of analytical methodologies directly contributes to the

safe and efficient operation of chemical plants and real-time analysis, in-process monitoring can control the formation of hazardous substances.
12. *Inherently Safer Chemistry for Accident Prevention:* Substances and chemical process should be chosen to minimize the potential for chemical accidents.

Classification of Green Solvents

Traditionally green solvents can be classified in five categories (Table 2):

1. Water
2. Ionic liquids
3. Fluorous solvents
4. Supercritical and subcritical fluids
5. Renewable solvents

Microbial infections are one of the most challenging and gradually persisting clinical complications worldwide. The widespread awareness and interest has been observed in past decade. Since the advent of green solvents, their potential applications, environmental effects, toxicity, antimicrobial activity, etc. are being studied and reported [7]. Based on a few recent publications, green solvent also made significant contributions with antimicrobial properties which suggest the potential therapeutic and preventive applications of it [35, 36]. Thus, an attempt is made to provide guidance on both organic and green solvent and to discuss and evaluate the antimicrobial susceptibility involving both the solvents.

Table 2. Classification of green solvents

Category	Definition	Advantages	Disadvantages	Example
Water	The most cheapest and common molecule on earth	Most used solvent, naturally occurring, abundant, non-toxic, low solubility of oxygen, environmental friendly	Not much suitable for organic synthesis, cumbersome purification, non-volatility leads to difficulty in removal, poor solvent for lipophilic compounds, high energy consumption in downstream processing	Water
Ionic liquids	Salts with high melting points below 100°C	Can be used as alternative solvents, good conductor of electricity, low vapor pressure and thermally stable, solubilizes organic, inorganic and even metals	Lengthy synthesis make it economic inefficient, limited data regarding health, safety and environment hazards	Ethylammonium nitrate, Monoethanolamine, deep eutectic solvents
Supercritical Solvent	A critical point of a solvent representing the highest temperature and pressure at which the substance can exists as vapor and liquid in equilibrium.	Non-toxic, Non-explosive, readily available, easy removable from the product, conventional critical properties, adjustable density, used as biocatalyst dissolve non-polar and slightly polar compounds	Requires high pressure, expensive setups, heat transfer limitations, non-polar, weak solvent	sCO_2, sH_2O

Table 2. (Continued)

Category	Definition	Advantages	Disadvantages	Example
Fluorous solvents	The highly fluorinated solvents used as catalyst and in separation process	Chemical and thermal stability, temperature dependent miscibility with conventional organic solvents	High persistence and stability which negatively acts in it degradation, highly corrosive and hard, highly expensive, accumulates in the atmosphere and can remain up to thousand years	Teflon, Perfluorooctanoic acid (PFOA), Perfluoromethylcyclohexane (PFCH), Perfluorooctanesulfonic acid (PFOS), Fluor surfactant Perfluorohexane FC-72, etc.
Renewable solvents	Solvents obtained from natural and renewable sources	Biologically resourced lower toxicity, low vapor pressure, low cost, non-ozone depleting compounds	Atmospheric pollution, flammability, not carbon-neutral	Glycerol, Glutamic acid, Sorbitol, Fumaric acid, Succinic acid, etc.

PHYTOCHEMICAL EXTRACTION FROM PLANTS USING ORGANIC SOLVENT VS. GREEN SOLVENT

Results of phytochemical screening were evaluated concerning extracts of both green and organic solvent from various medicinal plants. The efficiency of these solvents can be predicted by comparing the various studies revealing a wide range of phytoconstituents in medicinal plants. The three common and major phytochemicals groups are alkaloids, phenols and flavonoids will be discussed broadly with respect to its extraction from organic and green solvent.

Alkaloids

Alkaloids are the widest group of secondary metabolites having alkali-like properties primarily composed of the nitrogen atom in its heterocyclic ring structure, extracted from the plants, animals and microbial sources having efficient medicinal properties. One of the studies on *Peumusboldus* leaves, a medicinal tree, identified the characteristic alkaloids concerning the extraction from both of its organic (methanol, ethanol, water, etc.) and green solvent. In comparison with the efficiency, green solvents were found eight times more efficient than methanolic extraction [37]. Jiang and co-workers had the same observation as superiority in efficient alkaloid extraction of DESs (green solvent) over conventional methanol extraction was recorded [38]. With more such studies on comparison, Takla and coworker's comprehensively evaluated the potential and effectiveness of green solvents (NADES and Ionic surfactants) and conventional solvent which revealed that NADES and surfactants were significantly more efficient in alkaloid extraction than the conventional solvents and water [39], and same observations were made with Shwanky and other researchers [40-42].

Phenols

Phenols are the largest and most abundant group of phytochemicals, present in almost all plants. It helps in the process of growth, reproduction, act against pathogens and predators with anti-inflammatory, anti-microbial, anti-cancerous, etc. Green solvents have advantage of being biodegradable, less toxic and easy handling. Even, many researches based on them prove the efficiency and potential of green solvents over conventional solvents. Ruesgas-Ramón and co-fellows recorded the best yield of phenolic acids were higher in green solvents than methanolic acid. Similar study in 2016 by Peng and fellows also evaluated the same results [43, 44]. Study of phenolic compound in same Boldus leaves concludes 22 phenolic compounds from the methanolic extract which were similar to the results

Simirgiotis and coworkers, who identified 52 phenolic compounds in water extract. When compared with the efficiency of extraction method, green solvent was more efficient than conventional one [37, 45-48].

Flavonoids

Flavonoids are the most important class of secondary metabolites having polyphenolic structure. They are widely present in vegetables, fruits and beverages. They are favorable in many diseases like cancer, Alzheimer's, atherosclerosis, etc. [49]. Extraction efficiency of flavonoids from green solvents also tends to be higher as compared to organic or conventional solvents. Many studies and researches on various green solvent extractions support the data on efficiency and potentiality of it [44, 50-54]. Peng and others, in 2016 evaluated the flavonoid extraction potential of green solvent from five Chinese herbal medicines [44]. The study concluded the higher efficiency of green solvent than methanolic extracts. Significant increase in flavonoid content was noted by Bi and co-fellows, where Flavonoid was found 8 folds higher than water extract and 1.06 folds higher than methanolic extracts [50]. Zhuang also conferred similar results when extraction was analyzed using HPLC method, where green solvents extracted almost 90% of flavonoids as compared to methanol and then water [51], and same results were observed in other studies too [52, 53].

Antimicrobial Activity

The antimicrobial property of medicinal plants is known since antiquity. Many studies have been conducted to evaluate the antimicrobial activity of medicinal plants in different solvent system. Though past 2-3 decades have been a major breakthrough in antimicrobial compounds, microbes are evolving increasingly strong to antibiotic, while a pharmaceutical revolution is still needed. Several studies have been carried out to assess these organic and green solvents for their antimicrobial activity. Silva et al. has analyzed that no drug resistance was recorded on saturated fatty acids but combination with other extracts forming green solvents significant antimicrobial action towards gram positive strain and *C. albicans* [54]. Zhao too has similar results where amine, alcohol and sugar based green solvent extracts from *Saphora japonica* showed no resistance but significant results were seen in combinations with organic acids against four bacterial strains [55]. Similar study on ionic liquids showed significant microbial activities on various strains of bacteria and fungi and these studies were further supported by other researches (56-58). Evaluation of the antimicrobial activity of garlic extract delivered in organic solvent (isopropanol) showed dose-dependent antimicrobial activity against the tested pathogenic bacteria (Gram negative and positive) and *Aspergillus spp.* On the other hand, water-based micro emulsion showed better antimicrobial activity than emulsions at the same concentration [56].

CONCLUSION

This chapter provides insights into the analysis of organic and green solvents along with their biomedical properties. The study summarizes the importance, potential, and efficacy of green solvents. Green chemistry was introduced in the mid-late 90s, to replace the already present hazardous organic solvent by promoting innovative research with sustainable technology development. Owing to its properties of easy availability, less volatile, non-toxic, lesser amount, easy preparation, etc. the green solvent has enormous potential to replace the conventional methods. The main aim of the article is to understand the potency of extraction solvent and its antimicrobial actions. This review pointed out that green solvents are more potent than organic solvents in terms of extraction potential of phytochemicals and the antimicrobial activity was significantly higher in green solvents than organic. The efficient extraction of phytochemicals by "green" solvents provides an eco-viable alternative of the conventional solvents. These green solvents even can improve the efficiency of the extraction. Heat-labile phytochemicals can be extracted without degradation. Lesser non-toxicity of these solvents makes chemical processes economically as well as ecologically sustainable. More investigation is needed to be conducted in this area before taking it to the industrial platform.

REFERENCES

[1] Płotka-Wasylka, J., Rutkowska, M., Owczarek, K., Tobiszewski, M. and Namieśnik, J. (2017). Extraction with environmentally friendly solvents. *TrAC Trends in Analytical Chemistry*, 91, 12-25.

[2] Yadav, R. N. S. and Agrawala, M. (2011). Phytochemical analysis of some medicinal plants. *Journal of Phytology*, 3, 10-14.

[3] Agbafor, K. N. and Nwachukwu, N. (2011). Phytochemical analysis and antioxidant property of leaf extracts of *Vitexdoniana*and *Mucunapruriens*. *Biochemistry Research International*, 2011, 1-4.

[4] Zohra, S. F., Meriem, B., Samira, S. and Alsayadi-Muneer, M. (2012). Phytochemical screening and identification of some compounds from mallow. *Journal of Natural Product Plant Resource*, 2, 512-516.

[5] Iloki-Assanga, S. B., Lewis-Lujan, L. M., Lara-Espinoza, C. L., Gil-Salido, A. A., Fernandez-Angulo, D., Rubio-Pino, J. L. and Haines, D. D. (2015). Solvent effects on phytochemical constituent profiles and antioxidant activities, using four different extraction formulations for analysis of *Bucidabuceras* L. and *Phoradendroncalifornicum*. *BMC Research Notes*, 8, 396.

[6] Mazandarani, M., Zarghami, P., Zolfaghari, M. R., Ghaemi, E. A. and Bayat, H. (2012). Effects of solvent type on phenolics and flavonoids content and antioxidant activities in *Onosmadichroanthum*Boiss. *Journal of Medicinal Plants Research*, 6,4481–4488.

[7] Tiwari, J., Jain, V., Ratre, Y. K., Mishra, P. K. and Patra, P. K. (2018). In vitro Antioxidative and antifungal assessment of natural products against causative agents of Otomycosis and Candidiasis. *Indian Journal of Agricultural Biochemistry*, 31, 17-24.

[8] https://www.who.int/news-room/feature-stories/ten-threats-to-global-health-in-2019.

[9] Pollet, P., Davey, E. A., Urena-Benavides, E. E., Eckert, C. A. and Liotta, C. L. (2014). Solvents for sustainable chemical processes. *Green Chemistry*, 16, 1034-1055.

[10] Ngo, T. V., Van, T., Scarlett, C. J., Bowyer, M. C., Ngo, P. D. and Vuong, Q. V. (2017). Impact of different extraction solvents on bioactive compounds and antioxidant capacity from the root of *Salaciachinensis*L. *Journal of Food Quality*, 2017, 1-8.

[11] Ballesteros-Gomez, A., Sicilia, M.D. and Rubio, S. (2010). Supramolecular solvents in the extraction of organic compounds. A review. *Analytica Chimica Acta*, 677, 108-130.

[12] Byrne, F. P., Jin, S., Paggiola, G., Petchey, T. H., Clark, J. H., Farmer, T. J. and Sherwood, J. (2016). Tools and techniques for solvent selection: green solvent selection guides. *Sustainable Chemical Processes*, 4, 7.

[13] Ashcroft, C.P., Dunn, P.J., Hayler, J.D. and Wells, A.S. (2015). Survey of solvent usage in papers published in organic process research & development 1997–2012. *Organic Process Research and Development*, 19, 740–747.

[14] Lock, E. A. (1990). DR Buhler and DJ Reed (eds) Ethel browning's toxicity and metabolism of industrial solvents, 2nd edn, Vol. 2, nitrogen and phosphorus solvents. Elsevier, Amsterdam, 1990; 493 pp., 375.00 DG. *Journal of Applied Toxicology*, 10(5), 389-389.

[15] Sicaire, A. G., Vian, M., Fine, F., Jofre, F., Carre, P., Tostain, S. and Chemat, F. (2015). Alternative bio-based solvents for extraction of fat and oils: solubility prediction, global yield, extraction kinetics, chemical composition and cost of manufacturing. *International Journal of Molecular Sciences*, 16, 8430–8453.

[16] Virot, M., Tomao, V., Ginies, C. and Chemat, F. (2008). Total lipid extraction of food using d-limonene as an alternative to n-hexane. *Chromatographia*, 68, 311–313.

[17] Gissi, A., Lombardo, A., Roncaglioni, A., Gadaleta, D., Mangiatordi, G.F., Nicolotti, O. and Benfenati, E. (2015). Evaluation and comparison of benchmark QSAR models to predict a relevant REACH

endpoint: the bioconcentration fact (BCF). *Environmental Research*, 137, 398–409.

[18] Tebby, C., Mombelli, E., Pandard, P and Pery, A.R. R. (2011). Exploring an ecotoxicity database with the OECD (Q)SAR Toolbox and DRAGON descriptors in order to prioritise testing on algae, daphnids, and ish. *Science of the Total Environment*, 409, 3334–3343.

[19] *World Health Organization* (2015), IARC monographs on the evaluation of carcinogenic risks to human. https://monographs.iarc.fr/cards_page/staff/.

[20] Spencer, P. S. and Schaumburg, H. H. (1985). Organic solvent neurotoxicity. *Scandinavian Journal of Work Environment Health*, 11, 53-60.

[21] Liang, Q., Newman, P.A., Daniel, J.S., Reimann, S., Hall, B.D., Dutton, G. and Kuijpers, L.J. (2014). Constraining the carbon tetrachloride (CCl4) budget using its global trend and interhemispheric gradient. *Geophysical Research Letters*, 28, 5307-15.

[22] Joshi, D. R. and Adhikari, N. (2019). An overview on common organic solvents and their toxicity. *Journal of Pharmaceutical Research International*, 28, 1-18.

[23] Yamamoto, S., Kasai, T., Matsumoto M., Nishizawa T., H. Arito, Nagano K., and Matsushima, T. (2002). Carcinogenicity and chronic toxicity in rats and mice exposed to chloroform by inhalation. *Journal of Occupational Health*, 44, 283-293.

[24] Branchplower, R. V., Nunn, D. S., Highet, R. J., Smith, J. H., Hook, J. B. and Pohl, L. R. (1984). Nephrotoxicity of cloroform: metabolism to phosgene by the mouse kidney. *Toxicology and Applied Pharmacology*, 72, 159-168.

[25] Hossaini, R., Chipperfield, M. P., Montzka, S. A., Rap, A., Dhomse, S. and Feng, W. (2015). Efficiency of short-lived halogens at influencing climate through depletion of stratospheric ozone. *Nature Geoscienc*, 8, 186-190.

[26] DeCaprio, A. P. and O'Neill E. A. (1985). Alteratios in rat axonal cytoskeleton proteins inducred by *in vitro* and *in vivo* 2,5-hexanedione exposure. *Toxicology and Applied Pharmacology*, 78, 235-247.

[27] Khan, M. A., Ahmad, R. and Srivatava, A. N. (2017). Effect of ethyl acetate aroma on viability of human breast cancer and normal kidney epithelial cells *in vitro*. *Integrative Medicine Research*, 6, 47-59.
[28] Clay, K. L., Murphy, R. C. and Watkins, W. D. (1975). Experimental methanol toxicity in the primates: analysis of metabolic acidosis. *Toxicology and Applied Pharmacology*, 34, 49-61.
[29] Tarczykowska, A., Lekow, K. Z. C., Farmaceutyczny,W., Bydgoszczy, W. C., Toruniu, U. M. K., Jurasza, A. and Bydgoszcz. (2017).Green solvents. *Journal of Education Health and Sport*, 7, 224-232.
[30] Aparicio, S. and Alcalde, R. (2009). The green solvent ethyl lactate: an experimental and theoretical characterization. *Green Chemistry*, 11, 65-78.
[31] Jutz, F., Adanson, J. M. and Balker, A. (2011). Ionic liquids and dense carbon dioxide: A beneficial biphasic system for catalysis. *Chemical Reviews*, 111, 322–353.
[32] Li Ch., J. and Trost B. M. (2008). Green chemistry for chemical synthesis. *Proceedings of the National Academy of Sciences of the United States of America*, 105, 13197-13202.
[33] Capello, C., Fischer, U. and Hungerbuhler, K. (2017). What is a green solvent? A comprehensive framework for the environmental assessment of solvents. *Green Chemistry*, 9, 927-934.
[34] Promila and Singh, S. (2018). Applications of green solvents in extraction of phytochemicals from medicinal plants: A review. *The Pharma Innovation Journal*, 7, 238-245.
[35] Radosevic, K., Bubalo, M. C., Srcek, V. G., Grgas, D., Dragicevic, T. L. and Redovnikovic, I. R. (2015). Evaluation of toxicity and biodegradability of choline chloride based deep eutectic solvents. *Ecotoxicology and Environmental Safety*, 1, 46-53.
[36] Radosevic, K., Canak, I., Panic, M., Markov, K., Bubalo, M. C., Frece, J., Srcek, V. G. and Redovniković, I. R. (2018). Antimicrobial, cytotoxic and antioxidative evaluation of natural deep eutectic solvents. *Environmental Science and Pollution Research*, 14, 14188-14196.

[37] Torres-Vega, J., Gómez-Alonso, S., Pérez-Navarro, J. and Pastene-Navarrete, E. (2020). Green extraction of alkaloids and polyphenols from *Peumusboldus* leaves with natural deep eutectic solvents and profiling by HPLC-PDA-IT-MS/MS and HPLC-QTOF-MS/MS. *Plants*, 9, 242.

[38] Jiang, Z. M., Wang, L. J., Gao Z., Zhuang, B., Yin, Q. and Liu, E. H. (2019). Green and efficient extraction of different types of bioactive alkaloids using deep eutectic solvents. *Microchemical Journal*, 145, 345-353.

[39] Takla S. S., Shawky E., Hammoda H. M. and Darwish F. A. (2018). Green techniques in comparison to conventional ones in the extraction of Amaryllidaceae alkaloids: Best solvents selection and parameters optimization. *Journal of Chromatography A*, 1567, 99-110.

[40] Fuentes-Barros G., Castro-Saavedra S., Liberona L., Acevedo-Fuentes W., Tirapegui C., Mattar C. and Cassels B. K. (2018). Variation of the alkaloid content of Peumusboldus (boldo). *Fitoterapia*, 127, 179-185.

[41] Shawky E., Takla S. S., Hammoda H. M. and Darwish F. A. (2018). Evaluation of the influence of green extraction solvents on the cytotoxic activities of Crinum (Amaryllidaeae) alkaloid extracts using in-vitro-in-silico approach. *Journal of Ethnopharmacology*, 227, 139-149.

[42] Fuentes-Barros G., Castro-Saavedra S., Liberona L., Acevedo-Fuentes W., Tirapegui C., Mattar C. and Cassels B. K. (2018). Variation of the alkaloid content of *Peumusboldus* (boldo). *Fitoterapia*, 127, 179-185.

[43] Ruesgas-Ramón M., Figueroa-Espinoza M.C. and Durand E. (2017). Application of deep eutectic solvents (DES) for phenolic compounds extraction: overview, challenges, and opportunities. *Journal of agricultural and food chemistry*, 65, 3591-3601.

[44] Peng, X., Duan, M. H., Yao, X. H., Zhang, Y. H., Zhao, C. J., Zu, Y. G. and Fu, Y. J. (2016). Green extraction of five target phenolic acids from *Loniceraejaponicae*Flos with deep eutectic solvent. *Separation and Purification Technology*, 157, 249-257.

[45] Simirgiotis, M. J. and Schmeda-Hirschmann, G. (2010). Direct identification of phenolic constituents in Boldo Folium

(*Peumusboldus* Mol.) infusions by high-performance liquid chromatography with diode array detection and electrospray ionization tandem mass spectrometry. *Journal of Chromatography A*, 1217(4), 443-449.

[46] Gu T., Zhang M., Tan T., Chen J., Li Z., Zhang, Q. and Qiu, H. (2014). Deep eutectic solvents as novel extraction media for phenolic compounds from model oil. *Chemical Communication*, 50, 11749–11752.

[47] Nagarajan J., WahHeng W., Galanakis C. M., NagasundaraRamanan R., Raghunandan, M. E., Sun J. and Prasad K. N. (2016). Extraction of phytochemicals using hydrotropic solvents. *Separation Science and Technology*, 51, 1151–1165.

[48] Yao X. H., Zhang D. Y., Duan M. H., Cui Q., Xu W. J., Luo M. and Fu Y. J. (2015). Preparation and determination of phenolic compounds from *Pyrolaincarnata*Fisch. with a green polyols based-deep eutectic solvent. *Separation and Purification Technology*, 149, 116–123.

[49] Panche, A. N., Diwan, A. D. and Chandra, S. R. (2016). Flavonoids: an overview. *Journal of Nutritional Science*, 5, e47.

[50] Bi, W., Tian, M. and Row, K. H. (2013). Evaluation of alcohol-based deep eutectic solvent in extraction and determination of flavonoids with response surface methodology optimization. *Journal of Chromatography A*, 1285, 22-30.

[51] Zhuang, B., Dou, L. L., Li, P. and Liu, E. H. (2017). Deep eutectic solvents as green media for extraction of flavonoid glycosides and aglycones from *PlatycladiCacumen*. *Journal of Pharmaceutical and Biomedical Analysis*, 134, 214-219.

[52] Dai, Y., Witkamp, G. J., Verpoorte, R. and Choi, Y. H. (2013). Natural deep eutectic solvents as a new extraction media for phenolic metabolites in *Carthamustinctorius* L. *Analytical Chemistry*, 85(13), 6272-6278.

[53] deBrito, E. S., de Araújo, M. C. P., Lin, L. Z. and Harnly, J. (2007). Determination of the flavonoid components of cashew apple (*Anacardiumoccidentale*) by LC-DAD-ESI/MS. *Food Chemistry*, 105(3), 1112-1118.

[54] Silva J. M., Silva E., Reis R. L. and Duarte A. R. (2019). A closer look in the antimicrobial properties of deep eutectic solvents based on fatty acids. *Sustainable Chemistry and Pharmacy*, 14,100192.

[55] Zhao, B. Y., Xu, P., Yang, F. X., Wu, H., Zong, M. H. and Lou, W. Y. (2015). Biocompatible deep eutectic solvents based on choline chloride: characterization and application to the extraction of rutin from *Sophora japonica*. *ACS Sustainable Chemistry and Engineering*, 3, 2746-2755.

[56] Pernak J., Sobaszkiewicz K. and Mirska I. (2011). Anti-microbial activities of ionic liquids. *Green Chemistry*, 5, 52-56.

[57] Silva J. M., Pereira C. V., Mano F., Silva E., Castro V., Sá-Nogueira I. and Duarte A. R. C. (2019). The therapeutic role of deep eutectic solvents based on menthol and saturated fatty acids on wound healing. *ACS Applied Bio Materials*,10, 4346-4355.

[58] El-Sayed H. S., Chizzola R., Ramadan A. A. and Edris A. E. (2017). Chemical composition and antimicrobial activity of garlic essential oils evaluated in organic solvent, emulsifying, and self-microemulsifying water based delivery systems. *Food Chemistry*, 221, 196–204.

In: A Comprehensive Guide ...
Editors: Silje A. Dahl et al.

ISBN: 978-1-53618-418-1
© 2020 Nova Science Publishers, Inc.

Chapter 5

PHYTOCHEMICAL ACTIVITY AGAINST DRUG-RESISTANT MICROBES: CURRENT STATUS AND FUTURE PROSPECTS

Md. Didaruzzaman Sohel[1]
and Andrew W. Taylor-Robinson[2],*

[1]Quality Assurance Department, Incepta Pharmaceuticals Ltd.,
Dhaka-1341, Bangladesh
[2]School of Health, Medical & Applied Sciences, Central Queensland
University, Brisbane, Australia

ABSTRACT

The advance of antimicrobial resistance to existing frontline therapeutics is widely recognized as a global health threat. In order to address the increasing challenge that this presents to patient treatment by the medical profession pharmaceutical and biotechnology sectors continue

---

* Corresponding Author's Address: Prof A.W. Taylor-Robinson, School of Health, Medical & Applied Sciences, Central Queensland University, 160 Ann Street, Brisbane, QLD 4000, Australia. Tel: +61 73295 1185; Email: a.taylor-robinson@cqu.edu.au.

to seek novel therapeutic agents to which pathogens, notably bacteria, are sensitive.

The discovery of antibiotics in the last century led to a rapid and profound reduction in morbidity and mortality associated with commonly occurring bacterial diseases. However, the ongoing heavy reliance and indiscriminate use has resulted, due to genetic mutation under selective pressure, in the emergence of antibiotic-resistant bacteria. In searching for new commercial sources of antimicrobials crude extracts of medicinal plants have attracted attention. The wide range of metabolites – for example, alkaloids, tannins and polyphenols – carry therapeutic potential as either novel antimicrobials or modifiers of existing resistance. Plant extracts containing such phytochemicals are able to bind to protein domains, thereby modifying or inhibiting protein-protein interactions. This enables these herbal derivatives to act as effective modulators of cellular metabolic pathways involved in the immune response, mitosis, apoptosis and signal transduction. Hence, the mechanism(s) of action may not necessarily be directly microbicidal but instead affect key events within the host cell that reduce the ability of bacteria, fungi and viruses to thrive in an intracellular environment.

In this chapter, a brief history of antibiotics and the spread of resistance is provided. We describe phytochemicals that are currently known and outline their antimicrobial activities. Future strategies to combat drug-resistant microbes are discussed.

Keywords: antibiotic; antimicrobial drug resistance; drug discovery; medicinal plant; phytochemical

INTRODUCTION

History of Antibiotics

Antibiotics, the wonder medicines of the 20th century, continue to play a vital role in treating bacterial infections. The three decades from the 1950s to the 1970s have become known as the golden age of discovery of novel antimicrobial agents. Penicillin, the first natural antibiotic, was found by chance by Alexander Fleming in 1928 [1, 2]. Antibiotics are drug remedies used to treat bacterial infections. The international medical community needs urgently to amend the way it prescribes and makes use of antibiotics.

Even if new drug treatments are developed, without behavioral changes by practitioners to manage better the expectations of patients seeking prescriptions, antibiotic resistance will remain a principal threat to global public health. These changes should also encompass a range of actions to reduce infections, such as increased rates of vaccination, improved hand washing techniques, practicing safe sex and hygienic food preparation [3]. The main classes of antibiotics are P-lactams, macrolides, glycopeptides, aminoglycosides, tetracyclines, lipopeptides, lincosamides, rifamycins, cationic peptides, phenicols, streptogramins, oxazolidinones, quinolones, pyrimidines and sulfonamides. The most common place of origin of these antimicrobial compounds is the phylum Actinobacteria of Gram-positive bacteria. Approximately 80% of actinobacterial-derived antibiotics are sourced from members of the genus *Streptomyces* [4, 5].

MECHANISMS OF ANTIBACTERIAL ACTIVITY

The antibacterial activity of an agent is attributed mainly to two mechanisms that encompass interfering chemically with the synthesis or function of important components of bacteria, and/or circumventing the traditional pathways of antibacterial resistance. There are multiple targets for the antibacterial agents that comprise: (I) bacterial protein biosynthesis; (II) bacterial cell wall biosynthesis; (III) bacterial cell membrane destruction; (IV) bacterial DNA replication and repair; and (V) inhibition of a metabolic pathway. In addition, microorganisms may show resistance to antibacterial agents through a variety of mechanisms. A traditional antibacterial mechanism is via inhibiting transpeptidase enzyme activity. This blocks the final cross-linking step, transpeptidation, in the synthesis of peptidoglycan. For this reason, the cell wall becomes porous and water enters the bacterium due to high internal osmotic pressure. Eventually, the cell becomes so swollen that it ruptures, causing instantaneous death.

Some bacterial species exhibit innate resistance to one or more classes of antimicrobial agents. In such cases, determining the mechanisms of antibacterial resistance will inform how best to hinder or prevent its development [6]. In particular, this involves activation of an efflux pump, disarming antibacterial agents via destruction of enzymes, attenuation of antibiotics by means of editing enzymes, and alteration of target structures inside the bacterium that have decreased affinity for antibacterial recognition [7]. The underlying reason for the spread of resistance to antibiotics by bacterial populations is the presence of plasmids that contain genetic material and which can replicate independently and be passed among bacterial cells of both the same and different species. Antibacterial agents show their activity by interfering with bacterial protein biosynthesis and cell wall biosynthesis, inhibiting nucleic acid synthesis and destroying bacterial membranes [8, 9].

INAPPROPRIATE USE OF ANTIBIOTICS AND CAUSES OF RESISTANCE

The misuse of antibiotics is a significant contributing factor in the emergence of bacterial drug resistance. With increasing patient movement and transglobal air travel, transmission of drug-resistant organisms from one country to another has escalated [10, 11]. In fact, the World Health Organization (WHO) has named antibiotic resistance as one of the most important public health threats of the 21st century [12].

Multidrug resistance in bacteria takes place via accumulation of resistance (R) plasmids, transposons or genes that each code for resistance to a selected agent, and/or due to the action of multidrug efflux pumps (EP) [13]. In addition, emergence and dissemination of multidrug resistance (MDR) pressure in pathogenic bacteria has emerged as a considerable public fitness risk. Consequently, there are fewer, or even on occasion, no frontline antimicrobial agents available to combat infections due to pathogenic bacteria [14-16]. Antibiotic resistance is becoming such a critical public health concern that the search for new antimicrobial molecules has extended from natural resources to phytochemicals. In recent years, there an increasing variety of studies has aimed to discover novel bioactive compounds derived from plants or phytochemicals with the hope to manipulate antibiotic-resistant microorganisms [17]. Antibacterial resistance may be defined as the resistance of bacteria to treatment with antibiotic agents that were originally discovered to be powerful for its decontamination. As antibiotics become ineffective towards resistant microorganism clinical infections persist in patients, placing them at increased risk of deteriorating health and even death [18].

Bacteria are capable of adaptation to fluctuating nutrient availability, undesirable environmental conditions, the presence of antimicrobial products as well as to host immune responses. Antibiotic-resistant bacteria are recognized with increasing frequency and therefore new antimicrobials are needed to manage those pathogens. Serious bacterial infections that have become resistant to generally used antibiotics have emerged as a major worldwide healthcare problem. It could be argued that the development by bacteria of resistance, such as MDR, is inevitable inasmuch as it represents a selected aspect of overall microbial evolution MDR, along with other bacterial resistance mechanisms, make conventional treatment of bacterial infections and the management of biofilms produced by pathogenic bacteria increasingly ineffective [19]. With the progressively limited performance of frontline antibiotics, there is a pressing need to identify alternative sources of antimicrobial agents. Into this infection control void comes the potential to harness phytochemicals.

MECHANISMS OF ANTIMICROBIAL RESISTANCE

Mechanisms of drug resistance fall into several categories including drug inactivation/alteration, modification of drug-binding sites/targets, changes in cell permeability resulting in reduced intracellular drug accumulation, and biofilm formation [20-23]. Within this broad framework there are four main mechanisms of antimicrobial resistance that are currently recognized: (1) limiting uptake of a drug; (2) modifying a drug target; (3) inactivating a drug; (4) active efflux of a drug [24] (Figure 1). There is a natural variation in the capability of different bacteria to limit the uptake of antimicrobial agents. The structure and functions of the lipopolysaccharide layer of Gram-negative bacteria provides a barrier to certain categories of molecule. This gives those bacteria innate resistance to particular groups of large antimicrobial agents.

Multiple types of bacterial cell component may provide a suitable target for antimicrobial agents and there are just as many targets that may be modified by bacteria to enable resistance to those drugs. However, one mechanism of resistance to the β-lactam drugs that is exploited almost exclusively by Gram-positive bacteria is via alteration to the structure and/or number of penicillin-binding proteins (PBPs). There are two principal ways in which bacteria inactivate antimicrobial drugs; by actual degradation of the drug, or by transfer of a chemical group to the drug.

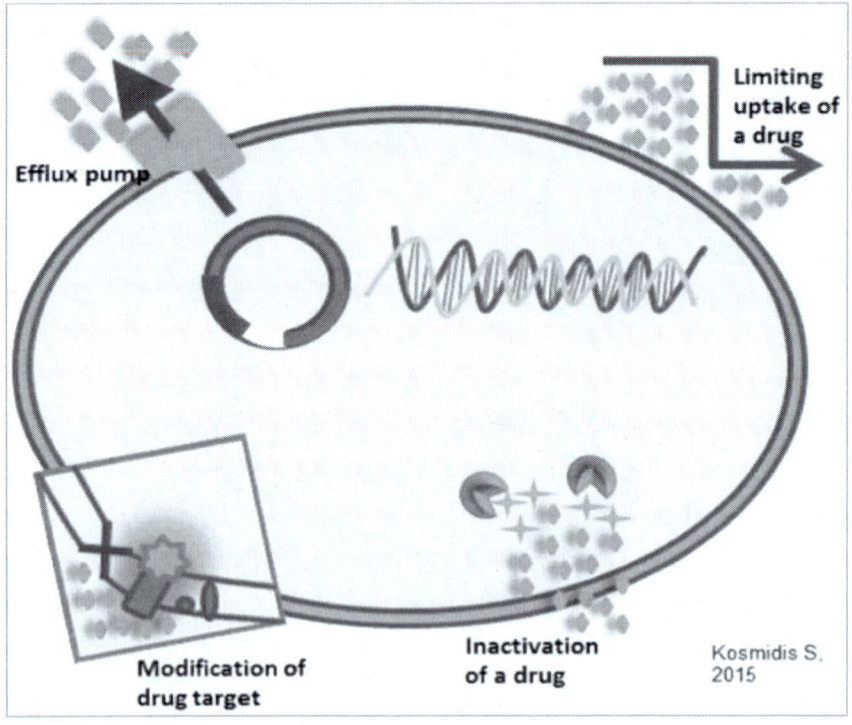

Figure 1. Antimicrobial resistance mechanisms [24].

The most widely used group of antimicrobial agents are the β-lactam drugs. The members of this group all share a specific core structure that consists of a four-sided β-lactam ring. Resistance to β-lactam drugs can occur via one of three general mechanisms: (1) preventing the interaction between the target PBP and the drug, usually by modifying the ability of the drug to bind to the PBP (this is mediated by alterations to existing PBPs or acquisition of other PBPs); (2) the presence of efflux pumps that can extrude β-lactam drugs; (3) hydrolysis of the drug by β-lactamase enzymes [24] (Figure 2). The β-lactamases are a very large group of drug-hydrolyzing enzymes. The tetracyclines are another important class of antibiotic with broad clinical utility that can be inactivated by hydrolyzation, the mechanism of which is under the control of the *tetX* gene.

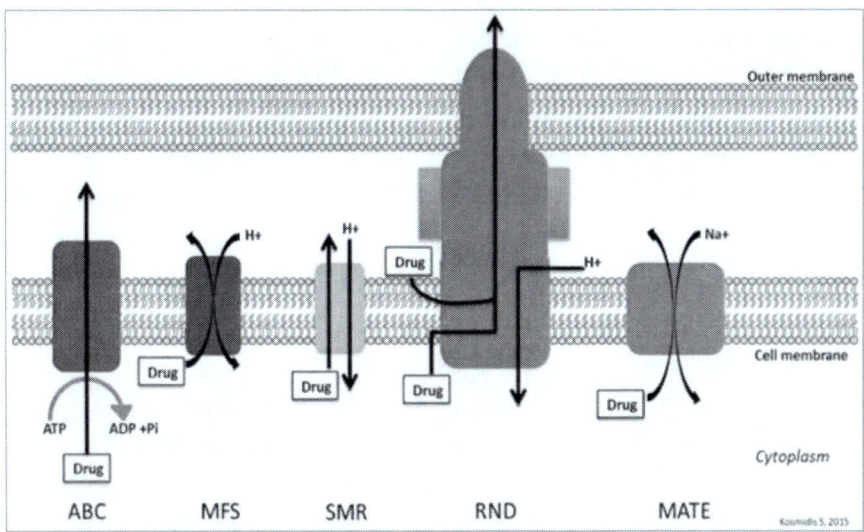

Abbreviations: ATP-binding cassette (ABC), Major facilitator superfamily (MFS), Small multidrug resistance (SMR), Resistance-nodulation-cell division (RND), Multidrug and toxic compound extrusion (MATE).

Figure 2. General structure of main efflux pump families [24].

ANTIMICROBIAL ACTIVITY OF MEDICINAL PLANTS

Plants are a primary supply of drugs and since historical times they have provided an alternative source of medication for treating all types of illnesses. Interestingly, around half of all pharmaceuticals approved by the US Federal Drug Administration have a plant origin. Only a few of these are used as antimicrobials for the reason that microbial properties are relied upon extensively. Since the widespread arrival of antibiotics in the 1950s the use of plant derivatives as antimicrobials has been virtually non-existent [25].

Natural products, especially those derived from microorganisms, have furnished the pharmaceutical industry with some of its most valuable assets, namely lead products in the search for novel antimicrobials. However, flora can also provide vital sources of antimicrobials and have been used for hundreds of years traditionally [26]. Natural products produced by vegetation, fungi, bacteria, insects and vertebrate animals have been isolated as biologically active pharmacophores. Today, approximately one-third of the top-selling drugs internationally are classified as either natural products or their derivatives from an ethnopharmacological background. According to the WHO, between 70–95% of the global human population relies on conventional drug treatments for their primary healthcare needs [27]. An extensive range of medicinal plants has been identified as a precious natural resource of potential antimicrobial compounds that may be exploited as an alternative therapy to treat problematic and persistent bacterial infections [28].

The search for the latest generation of antibiotics is a very expensive and time-consuming process, typically requiring about ten years to bring a newly identified antibiotic to commercial production. Hence, the quest for antibacterial substances derived from naturally occurring phytochemicals has gained growing significance alongside the exploration for new synthetic chemical compounds with antibiotic properties [29].

Phytochemicals form a large grouping of chemicals that are naturally present in vegetation, conferring color, flavor, aroma and texture. These compounds have evolved to innately defend plant species from the damaging effects of free radicals, viruses, bacteria and fungi. They are distributed widely in fruits, vegetables, legumes, entire grains, nuts, seeds, herbs and spices and in plant-based beverages including wine and tea. Phytochemicals can belong to several predominant categories based primarily on their chemical structure: alkaloids; sulfur-containing phytochemicals; terpenoids; and polyphenols [16].

Evaluating the antimicrobial activity of natural products in the laboratory, two different qualitative methods have been used: the agar diffusion test, employing two types of reservoirs (filter paper disc impregnated with compound-test and wells in dishes), and the bioautographic method (agar diffusion and chromatogram layer). The microdilution method is used for the determination of minimum inhibitory concentration (MIC). Investigating the effects of natural products on test bacteria, selected assays include indicator solution for determination of bacterial growth, agar diffusion well-variant, agar diffusion disc-variant, bioautographic method direct-variant (chromatogram layer), bioautographic method indirect-variant (agar diffusion) and MIC determination [30].

Antimicrobial activity has been identified in a number of medicinal plants such as *Combertum molle, Xanthium strumarium, Measalanceolata, Dodonae angustifolia, Asparagus africanus, Aloe barbarae Dyer, Cissus quadrangularis L., Jatropha zeyheri Sond, Ziziphus mucronata Willd., Datura stramonium, Lantana camara, Gentiana punctata L., Peltophorum africanum Sond., C. macrostachyus, C. aurea, Phyllanthus discoideus (Baill.) Muel-Arg., Strychnos spinose, Pterospermum acerifolium* and *Madhuca longifolia* [31]. Screening of such plants as a potential source of alternative antimicrobial agents is now being conducted all over the world. Antimicrobial properties of plants are attributed to the presence of active compounds, e.g., quinones, phenols, alkaloids, flavonoids, terpenoids, essential oils, tannins, lignans, glucosinolates and certain secondary metabolites. Other antimicrobial actions of plants include defense systems formed by peptides that are similar to human antimicrobial peptides in structure and function [32].

Traditionally, the crude extracts of different parts of medicinal plants, including root, stem, leaf, flower and fruit, were widely used for herbal treatments of a variety of human diseases. Medicinal plants contain several phytochemicals, including flavonoids, alkaloids, tannins and terpenoids, which possess antimicrobial properties. The antimicrobial activities of some plant species have been researched in detail. For example, the crude extracts of cinnamon, garlic, basil, curry, ginger, sage, mustard and other herbs exhibit antimicrobial properties against a wide range of Gram-positive and Gram-negative bacteria [33].

PATHOGENS PRIORITISED BY THE WHO FOR RESEARCH AND DEVELOPMENT OF NEW ANTIBIOTICS

The WHO recently published a global priority list of antibiotic-resistant bacteria in order to guide research, discovery and development of new antibiotics [34]. The remit of the report was to identify the most important resistant bacteria on a global level for which there is an urgent need to find novel treatments.

Priority 1: Critical
1. *Acinetobacter baumannii*, carbapenem-resistant
2. *Pseudomonas aeruginosa*, carbapenem-resistant
3. Enterobacteriaceae, carbapenem-resistant, ESBL-producing

Priority 2: High
1. *Enterococcus faecium*, vancomycin-resistant
2. *Staphylococcus aureus*, methicillin-resistant, vancomycin-intermediate and resistant
3. *Helicobacter pylori*, clarithromycin-resistant
4. *Campylobacter* spp., fluoroquinolone-resistant
5. Salmonellae, fluoroquinolone-resistant
6. *Neisseria gonorrhoeae*, cephalosporin-resistant, fluoroquinolone-resistant

Priority 3: Medium
1. *Streptococcus pneumoniae*, penicillin-non-susceptible
2. *Haemophilus influenzae*, ampicillin-resistant
3. *Shigella* spp., fluoroquinolone-resistant

DISCUSSION

Used in traditional remedies since ancient times natural products contain biologically active molecules that might act as important tools in disease therapy. Many pathological conditions have been treated using plant-derived medicines. Drug discovery is a multidimensional objective requiring evaluation of several parameters of both natural and synthetic compounds including pharmacokinetics, safety and efficacy during successive stages of drug candidate selection. The advent of cutting edge applications such as artificial intelligence and 'organ-on chip' and microfluidics technology that enhance drug conformational analysis means that automation has become an integrated aspect of the drug discovery process. This has resulted in an accelerated pace of identification and evaluation of candidate compounds whilst also facilitating novel ways of drug design and synthesis based on natural compounds.

Recent advances in analytical and computational techniques have opened new avenues to process complex herbal products and to examine their molecular structures in order to derive innovative and efficacious drugs. Predictive computational software has contributed to the creation of synthetic derivatives of natural products. The use of quantum computing, computational software and databases in modelling molecular interactions and in predicting capabilities and parameters needed for drug development, along with pharmacokinetic and pharmacodynamics, will result in more accurate and successful leads during the next era of drug advancement, especially to combat multidrug-resistant bacteria [35-37].

FUTURE PROSPECTS

Growing concern surrounding antibiotic resistance by bacteria pathogenic to humans is driving the urgent modification of existing antibiotics and the parallel development of new generation antibiotics. There are three commonly recognized pathways to fight against antibiotic resistance; namely antimicrobial chemicals derived from natural products, chemical compounds modified to become antibiotics, and phages. Antimicrobial compounds extracted from natural products have drawbacks concerning their isolation and purification. Phytochemicals isolated from natural products and then chemically synthesized via adjustments are likely to provide effective antimicrobial drugs in the near future. In this period of advancing clinical technology, modern drug discovery that draws on natural products will potentially increase the success rate of both finding and exploiting novel, potentially therapeutic moieties.

Natural product drug discovery stands as a major contributor to addressing the current demanding global health situation and to achieving sustainable development goals on public health. Over the coming years, the opportunity to exploit medicinal plant life as antimicrobial agents require an applied commercial focus that builds on basic research on phytochemical discovery. The isolation of biologically active compounds should be undertaken in light of the recognized properties of the plant and that will guide isolation principles. Thus, whilst fractions may be inferior to the entire extract, instead of invalidating the results, this process should aim to confirm the known antimicrobial properties of the plant. In summary, it is imperative to uncover new classes of antibiotic or antimicrobial agent with exclusive modes of action towards microbial pathogens that exhibit MDR [4].

CONCLUSION

Combinational drug use is greatly relied upon to treat bacterial contamination, yet even this treatment regimen may not prevent the emergence and spread of resistance among microbial pathogens. In order to overcome the challenges of antibiotic resistance, antimicrobial compounds that have a completely novel mechanism of action must be urgently sought. Those natural products are then required to undergo rigorous toxicological screening in preclinical and clinical trials prior to declaring that the identified drugs are safe for human use.

CONFLICT OF INTEREST STATEMENT

The authors declare that there are no conflicts of interest regarding to the publication of this chapter.

FUNDING

No funding was associated with the preparation of this chapter.

AUTHOR CONTRIBUTIONS

ATR proposed the chapter and supervised its preparation. MDS collated and reviewed literature. Each author contributed to writing the text, read and approved submission of the final manuscript.

REFERENCES

[1] Peterson E and Kaur P (2018). Antibiotic resistance mechanisms in bacteria: relationships between resistance determinants of antibiotic producers, environmental bacteria, and clinical pathogens. *Front. Microbiol.* 9: 2928. http://doi.10.3389/fmicb.2018.02928.

[2] Gould K (2016). Antibiotics: from prehistory to the present day. *J. Antimicrob. Chemother.* 71: 572-575. https://doi.org/10.1093/jac/dkv 484.

[3] World Health Organization (2018). *Antibiotic resistance fact sheet.* https://www.who.int/news-room/fact-sheets/detail/antibiotic-resistance. Last updated 5 February 2018.

[4] Aunjum A and Taylor-Robinson AW (2019). Phytochemicals produced to combat plant pathogens as a source of novel antimicrobial compounds against multidrug-resistant bacteria. *Curr. Top. Phytochem.* 15: 55-60. https://doi.org/10.1155/2019/5410923.

[5] Kapoor G, Saigal S and Elongavan A (2017). Action and resistance mechanisms of antibiotics: a guide for clinicians. *J. Anaesthesiol. Clin. Pharmacol.* 33: 300-305. https://dx.doi.org/10.4103%2Fjoacp.JOACP _349_15.

[6] Walsh C (2000). Molecular mechanisms that confer antibacterial drug resistance. *Nature* 406: 775-781. https://doi.org/10.1038/3502 1219.

[7] Khameneh B, Diab R, Ghazvini K and Fazly Bazzaz BS (2016). Breakthroughs in bacterial resistance mechanisms and the potential ways to combat them. *Microb. Pathog.* 95: 32-42. https://doi.org/ 10.1016/j.micpath.2016.02.009.

[8] Khameneh B, Iranshahy M, Soheili V and Bazzaz BSF (2019). Review on plant antimicrobials: a mechanistic viewpoint. *Antimicrob. Resist. Infect. Control* 8: 118. https://doi.org/10.1186/ s13756-019-0559-6.

[9] Sahraei S, Mohkami Z, Golshani F, et al., (2014). Antibacterial activity of five medicinal plant extracts against some human bacteria. *Eur. J. Exp. Biol.* 4: 194-196. http://www.imedpub.com/articles/antibacterial-activity-of-five-medicinal-plant-extracts-against-some-human-bacteria.pdf.

[10] Gupta PD and Birdi TJ (2017). Development of botanicals to combat antibiotic resistance. *J. Ayurveda Integr. Med.* 8 (4): 266–275. http://doi:10.1016/j.jaim.2017.05.004.

[11] Aminov RI (2010). A brief history of the antibiotic era: lessons learned and challenges for the future. *Front. Microbiol.* 1: 134. https://dx.doi.org/10.3389%2Ffmicb.2010.00134.

[12] World Health Organization (2014). *Antimicrobial Resistance: Global Report on Surveillance 2014*. https://www.who.int/drugresistance/docu ments/surveillancereport/en/.

[13] Nikaido H (2009). Multidrug resistance in bacteria *Annu. Rev. Biochem.* 78: 119-146. https://doi.org/10.1146/annurev.biochem.78. 082907.145923.

[14] Giamarellou (2010). Multidrug-resistant Gram-negative bacteria: how to treat and for how long. *Int. J. Antimicrob. Agents* 36, Suppl. 2, S50-S54. http://doi:10.1016/j.ijantimicag.2010.11.014.

[15] Boucher HW, Talbot GH, Bradley JS et al., (2009). Bad bugs, no drugs: no ESKAPE! An update from the Infectious Diseases Society of America. *Clin. Infect. Dis.* 48: 1-12. http://doi:10.1086/595011.

[16] Bhatia R and Narain JP (2010). Growing challenge of antimicrobial resistance in the South-East Asia Region – are we losing the battle? *Indian J. Med. Res.* 132: 482-486. http://www.ijmr.org.in/text.asp? 2010/132/5/482/73313.

[17] Subramani R, Narayanasamy M and Feussner KD (2017). Plant-derived antimicrobials to fight against multi-drug-resistant human pathogens. *3 Biotech.* 7: 172. http://doi:10.1007/s13205-017-0848-9.

[18] Barbieri R, Coppo E, Marchese A, Daglia M, et al., (2017). Phytochemicals for human disease: an update on plant-derived compounds antibacterial activity. *Microbiol. Res.* 196: 4-68. http://doi:10.1016/j.micres.2016.12.003.

[19] Simões M, Lemos M and Simões LC (2012). Phytochemicals against drug-resistant microbes. In: Patra AK (ed.), *Dietary Phytochemicals and Microbes*, pp. 185-205, Springer, Dordrecht, Netherlands. http://doi:10.1007/978-94-007-3926-0_6.

[20] Wright GD (2005). Bacterial resistance to antibiotics: enzymatic degradation and modification. *Adv. Drug Deliv. Rev.* 57: 1451-1470. http://dx.doi.org/10.1016/j.addr.2005.04.002.

[21] Li XZ and Nikaido H (2004). Efflux-mediated drug resistance in bacteria. *Drugs* 64: 159-204. https://doi.org/10.2165/00003495-200464 020-00004.

[22] Wilson DN (2014). Ribosome-targeting antibiotics and mechanisms of bacterial resistance. *Nat. Rev. Microbiol.* 12: 35-48. https://doi.org/10.1038/nrmicro3155.

[23] Santajit S and Indrawattana N (2016). Mechanisms of antimicrobial resistance in ESKAPE pathogens. *BioMed Res. Int.* 2016: 2475067. http://dx.doi.org/10.1155/2016/2475067.

[24] Reygaert WC (2018). An overview of the antimicrobial resistance mechanisms of bacteria. *AIMS Microbiol.* 4: 482-501. https://dx.doi.org/10.3934%2Fmicrobiol.2018.3.482.

[25] Cowan MM (1999). Plant products as antimicrobial agents. *Clin. Microbiol. Rev.* 12: 564-582. https://doi.org/10.1128/CMR.12.4.564.

[26] Clardy J and Walsh C (2004). Lessons from natural molecules. *Nature* 432: 637-641. https://www.nature.com/articles/nature03194.

[27] World Health Organization (2011). *The World Medicines Situation 2011*, 3rd edition, Geneva. http://apps.who.int/medicinedocs/documents/s20054en/s20054en.pdf.

[28] Manandhar S, Luitel S and Dahal RK (2019). *In vitro* antimicrobial activity of some medicinal plants against human pathogenic bacteria. *J. Trop. Med.* 2019: 1895340. https://doi.org/10.1155/2019/1895340.

[29] Mandal SM, Roy A, Ghosh AK, et al., (2014). Challenges and future prospects of antibiotic therapy: from peptides to phages utilization. *Front. Pharmacol.* 5: 1-12. https://dx.doi.org/10.3389%2Ffphar.2014.00105.

[30] Valgas C, de Souza SM, Smânia EFA, et al., (2007). Screening methods to determine antibacterial activity of natural products. *Braz. J. Microbiol.* 38: 369-380. https://doi.org/10.1590/S1517-83822007000200034.

[31] Dilbato T, Begna F and Joshi RK (2019). Reviews on challenges, opportunities and future prospects of antimicrobial activities of medicinal plants: alternative solutions to combat antimicrobial resistance. *Int. J. Herb. Med.* 7: 10-18. http://www.florajournal.com/archives/2019/vol7issue4/PartA/7-2-22-461.pdf.

[32] Chandra H, Bishnoi P, Yadav A, et al., (2017). Antimicrobial resistance and the alternative resources with special emphasis on plant-

based antimicrobials – a review. *Plants* 6: 16. http://doi:10.3390/plants6020016.

[33] Gonelimali FD, Lin J, Miao W, et al., (2018). Antimicrobial properties and mechanism of action of some plant extracts against food pathogens and spoilage microorganisms. *Front. Microbiol.* 9: 1639. http://doi:10.3389/fmicb.2018.01639.

[34] World Health Organization (2017). *WHO publishes list of bacteria for which new antibiotics are urgently needed.* https://www.who.int/en/news-room/detail/27-02-2017-who-publishes-list-of-bacteria-for-which-new-antibiotics-are-urgently-needed.

[35] Liberati A, Altman DG, Tetzlaff J, et al., (2009). The PRISMA statement for reporting systematic reviews and meta-analyses of studies that evaluate health care interventions: explanation and elaboration. *PLoS Med.* 6: e1000100. https://doi.org/10.1371/journal.pmed.1000100.

[36] *Frontiers* Research Topic (2020). *Natural products as a tool to design new anti-MDR lead molecules.* https://www.frontiersin.org/research-topics/10477/natural-products-as-a-tool-to-design-new-anti-mdr-lead-molecules.

[37] Thomford NE, Senthebane DA, Rowe A, et al., (2018). Natural products for drug discovery in the 21st century. Innovations for novel drug discovery. *Int. J. Mol. Sci.* 19: 1578. https://doi.org/10.3390/ijms19061578.

INDEX

A

accessibility, x, 149, 154
acetic acid, 9, 14, 19, 21, 22, 24, 26, 27, 40, 41
acetonitrile, 7, 8, 9, 10, 11, 12, 13, 19, 20, 21, 22, 23, 24, 25, 26, 27, 38, 39, 40, 41, 42, 43, 44, 45, 46, 47, 50, 52, 53, 66
acid, 8, 9, 10, 11, 13, 14, 19, 20, 21, 22, 23, 24, 25, 26, 38, 39, 40, 41, 42, 43, 44, 45, 46, 47, 48, 50, 51, 52, 53, 66, 72, 74, 85, 96, 97, 98, 99, 100, 103, 104, 106, 107, 109, 110, 111, 112, 116, 120, 140, 148, 158, 159
active compound, 3, 59, 77, 104, 135, 177
age, 62, 101, 133, 151, 170
algae, viii, 91, 92, 95, 164
alkaloid(s), vii, ix, x, 3, 10, 16, 17, 20, 24, 28, 29, 43, 48, 56, 57, 72, 76, 78, 85, 130, 136, 137, 138, 139, 148, 150, 151, 158, 159, 166, 170, 177, 178
amine(s), 29, 152, 161
amino, 3, 16, 48, 57, 140, 143
amino acid, 3, 48, 57, 140, 143
ammonium, 19, 21, 23, 24, 25, 27, 39, 40, 41, 44, 46, 66

anthraquinone, ix, 130, 136, 137, 138, 139
antibacterial activity, 7, 142, 147, 171, 182, 183, 184
antibiotic, x, 160, 170, 172, 173, 175, 176, 178, 180, 181, 182, 184
antibiotic resistance, 171, 172, 180, 181, 182
antimicrobial activity, x, 112, 139, 146, 150, 156, 160, 161, 168, 177, 184
antimicrobial drug resistance, 170
antioxidant(s), vii, viii, ix, 7, 24, 71, 72, 73, 74, 78, 79, 84, 91, 101, 102, 103, 104, 105, 106, 107, 108, 109, 110, 111, 112, 119, 120, 121, 124, 125, 139, 141, 143, 144, 145, 146, 150, 162, 163
antisickling, v, ix, 129, 130, 131, 134, 139, 141, 143, 144, 145, 146, 147, 148
antitumor, 111, 126, 139
apoptosis, xi, 101, 109, 111, 170
assessment, 25, 27, 35, 62, 73, 79, 86, 88, 139, 162, 165
atherosclerosis, 101, 104, 110, 111, 160
atmospheric pressure chemical ionization (APCI), 23, 25, 27, 30, 67, 75
authentication, 17, 52, 53, 62

B

bacteria, x, 2, 112, 161, 170, 171, 172, 173, 174, 176, 177, 178, 179, 180, 181, 182, 183, 184, 185
bacterial drug resistance, 172
bacterial infection, 170, 173, 176
benefits, 3, 30, 66
beverages, 69, 130, 160, 177
biological activity/activities, 7, 29, 86, 143
biomarkers, 17, 52, 53, 62, 65
biosynthesis, 112, 116, 171, 172
blood, ix, 28, 129, 131, 133

C

cancer, vii, ix, 2, 3, 92, 101, 104, 111, 139, 160
cannabinoids, 38, 55, 82
cannabis, 38, 55, 82
capsule, 23, 29, 44, 45, 77, 86
carbon, 103, 146, 153, 158, 164, 165
carbon tetrachloride, 146, 153, 164
cardiac glycoside, 136, 137, 138
cardiovascular disease, ix, 92, 139
cell line, 108, 109, 111, 113
cetrarioid clade, viii, 91, 92, 93, 96, 97, 104, 106, 109, 112
challenges, 80, 141, 166, 181, 182, 184
chemical, viii, 2, 4, 5, 6, 14, 15, 16, 17, 20, 21, 22, 23, 26, 27, 28, 30, 35, 48, 50, 51, 55, 57, 59, 61, 67, 70, 71, 72, 74, 75, 77, 80, 84, 96, 120, 130, 131, 135, 139, 151, 152, 154, 155, 156, 162, 163, 165, 174, 176, 177, 180
chemicals, 3, 4, 36, 66, 154, 176, 180
chemometrics, v, vii, viii, 1, 2, 6, 8, 10, 16, 20, 21, 22, 23, 24, 25, 26, 32, 33, 35, 36, 38, 40, 41, 42, 43, 44, 46, 59, 66, 68, 70, 71, 73, 74
children, 131, 132, 133, 141

China, 34, 35, 93, 116
Chinese medicine, 30, 48, 55, 56, 59, 63, 70, 74
chloroform, 28, 151, 153, 164
chromatograms, 17, 30, 31, 49, 52, 53, 58
chromatographic technique, 66
chromatography, viii, 2, 4, 5, 6, 7, 13, 18, 37, 55, 63, 66, 67, 69, 70, 76, 82, 84, 85, 89, 95
classes, vii, x, 70, 150, 171, 180
commercial, x, 2, 7, 32, 60, 65, 73, 170, 176, 180
comparative analysis, x, 84, 150
complexity, 62, 66, 143
composition, ix, 6, 18, 33, 35, 36, 62, 71, 74, 78, 80, 92, 96, 97, 120, 146, 149, 152, 163, 168
compounds, vii, ix, 2, 4, 5, 6, 7, 8, 10, 11, 12, 13, 14, 16, 18, 20, 21, 22, 23, 24, 25, 26, 27, 28, 29, 30, 32, 33, 34, 35, 36, 37, 38, 39, 40, 41, 42, 43, 44, 46, 47, 48, 50, 51, 55, 56, 57, 58, 59, 60, 63, 64, 66, 69, 74, 76, 79, 82, 92, 95, 96, 97, 99, 100, 103, 104, 105, 106, 107, 113, 130, 131, 134, 140, 146, 150, 151, 152, 153, 154, 157, 158, 160, 162, 163, 171, 173, 176, 179, 180, 181, 182, 183
comprehensive two-dimensional liquid chromatography (LC x LC), 63, 64, 88
configuration, 13, 72, 87
constituents, 4, 20, 21, 22, 23, 26, 29, 30, 38, 47, 57, 60, 61, 73, 74, 75, 76, 77, 80, 81, 83, 84, 130, 142, 143, 151, 166
correlation(s), 47, 59, 79, 81, 107
cosmetic(s), 2, 50, 130
cost, 131, 152, 158, 163
coumarins, 39, 48, 58, 136, 138
crises, 133, 134, 140
cure, ix, 130, 140

D

data processing, viii, 2, 6, 86
database, 54, 87, 164
degradation, 92, 158, 162, 174, 183
dehydration, 131, 135, 139, 142
derivatives, xi, 12, 14, 15, 50, 170, 176, 179
detection, vii, 5, 6, 7, 8, 9, 10, 12, 13, 14, 16, 50, 52, 55, 60, 64, 66, 68, 73, 85, 167
detection system, 6, 7, 8, 16
diabetes, 101, 104, 110, 124, 125
discriminant analysis, 17, 62, 65
discrimination, ix, 17, 50, 51, 53, 83, 129
diseases, vii, ix, x, 2, 3, 48, 66, 92, 104, 108, 130, 132, 135, 139, 150, 151, 160, 170, 178
disorder, ix, 129, 130, 132, 141, 145
distribution, 3, 35, 79, 117
diversity, 3, 4, 69
DNA, 93, 95, 101, 108, 110, 111, 171
drug discovery, 69, 170, 179, 180, 185
drug resistance, 161, 170, 172, 173, 182, 183
drugs, vii, ix, 2, 3, 130, 131, 134, 149, 151, 174, 175, 176, 179, 180, 181, 183

E

efficiency, vii, x, 11, 27, 51, 52, 64, 87, 134, 150, 151, 154, 155, 158, 159, 160, 161, 164
electron(s), 29, 101, 105, 106
energy, 55, 56, 63, 155, 157
environment, xi, 3, 92, 153, 154, 155, 157, 170
enzyme(s), ix, 79, 92, 101, 102, 103, 108, 109, 110, 171, 172, 175
ESI, 20, 21, 22, 23, 24, 25, 26, 28, 29, 38, 39, 40, 41, 42, 43, 44, 45, 46, 47, 49, 51, 52, 53, 62, 79, 86, 87, 167
ester, 14, 15, 16, 140, 153

ethanol, 28, 43, 85, 119, 142, 143, 159
ethyl acetate, 48, 58, 151, 165
evaporative light scattering detection (ELSD), 7, 10, 16, 17, 66
exposure, ix, 149, 153, 154, 164
extraction, ix, 14, 50, 57, 80, 82, 86, 149, 150, 151, 153, 158, 159, 160, 161, 162, 163, 165, 166, 167, 168
extracts, x, 4, 12, 13, 14, 17, 28, 40, 49, 52, 57, 72, 74, 78, 80, 83, 84, 87, 104, 105, 107, 108, 119, 134, 139, 140, 141, 142, 143, 145, 146, 147, 148, 158, 160, 161, 162, 166, 170, 177, 182, 185

F

factor analysis (FA), 9, 16, 17
families, ix, 61, 65, 66, 93, 130, 135, 139, 175
flavonoids, vii, ix, x, 3, 7, 10, 24, 26, 28, 29, 35, 48, 50, 52, 57, 61, 64, 65, 76, 130, 135, 136, 137, 138, 139, 150, 151, 158, 160, 162, 167, 177, 178
flavonol, 12, 21, 32, 33, 76
flight, viii, 2, 5, 54, 76, 77, 82, 88
fluorescence, 5, 6, 14, 66, 106
fluorescence (FL) detection, 10, 14
food, 13, 50, 69, 70, 82, 87, 163, 166, 171, 185
formation, 130, 156, 173
formula, 32, 74, 79
free radicals, 101, 104, 107, 139, 177
fruits, 4, 77, 140, 143, 160, 177
fungus/fungi, vii, viii, xi, 2, 91, 92, 93, 95, 161, 170, 176, 177

G

gene therapy, 134, 140, 142
genes, 92, 130, 172
genus, 93, 94, 105, 171

ginseng, 52, 61, 62, 65
glucoside, 32, 33, 34
glucosinolates, 32, 41, 44, 47, 81, 177
glutathione, 102, 103, 108, 109, 110
green solvent, x, 149, 150, 154, 156, 157, 158, 159, 160, 161, 163, 165
growth, 3, 35, 80, 112, 133, 159, 177

H

health, x, 36, 131, 132, 151, 152, 153, 154, 157, 162, 169, 173, 180, 185
hemoglobin, 110, 130, 134, 147
herbal, xi, 26, 27, 28, 29, 30, 31, 35, 42, 46, 62, 70, 79, 84, 87, 130, 140, 145, 148, 160, 170, 177, 179
herbal medicine, 26, 28, 35, 42, 70, 84, 87, 160
hierarchical cluster analysis (HCA), 8, 9, 10, 16, 22, 26, 35, 67
high-resolution mass spectrometry, viii, 2, 5, 67, 82, 84, 85
high-speed counter-current chromatography (HSCCC), 56
history, vii, xi, 151, 170, 182
host, viii, xi, 91, 133, 170, 173
HPLC-UV, 7, 11, 12, 13, 17, 71, 73
human, 108, 109, 110, 111, 134, 154, 155, 164, 165, 176, 177, 178, 181, 182, 183, 184
hybrid, 5, 18, 36, 38, 39, 40, 41, 42, 43, 44, 45, 46, 47, 54, 55, 67, 84, 85, 86, 87
hydrogen, 101, 102, 105, 106, 108, 109, 111, 112
hydrogen peroxide, 101, 102, 108, 109, 111, 112
hydroxyl, 101, 103, 104, 107, 110

I

identification, viii, 2, 4, 5, 13, 30, 35, 36, 48, 49, 51, 52, 53, 54, 55, 56, 57, 59, 62, 63, 65, 66, 67, 68, 74, 75, 78, 82, 84, 87, 88, 95, 131, 162, 166, 179
immune response, xi, 170, 173
in vitro, vii, ix, 59, 73, 78, 82, 104, 105, 111, 113, 129, 147, 164, 165
India, 69, 93, 113, 114, 117, 129, 131, 132, 142, 149
industry/industries, 4, 151, 152, 154, 176
infection, 133, 134, 150, 173
inhibition, 73, 105, 112, 116, 131, 140, 171
ion mobility mass spectrometry (IMS), 36, 64, 65, 68
ionization, 13, 27, 28, 29, 30, 50, 51, 52, 67, 75, 76, 78, 79, 87, 88, 167
ions, 32, 36, 48, 57, 63, 108
iron, 101, 108, 133, 134
isolation, viii, 2, 4, 35, 36, 95, 151, 180
isomers, 36, 48, 53, 57

K

kaempferol, 32, 33, 64
kidney(s), 102, 111, 164, 165

L

LC-MS, viii, 2, 5, 6, 17, 19, 29, 30, 33, 59, 64, 70, 77, 78, 79
LC-MS/MS, viii, 2, 5, 19, 29, 30, 78, 79
lead, 150, 154, 176, 185
lichen(s), v, vii, viii, ix, 39, 83, 91, 92, 93, 94, 95, 96, 97, 98, 99, 100, 104, 105, 106, 107, 108, 109, 110, 112, 113, 114, 115, 116, 117, 118, 119, 120, 121, 122, 123, 124, 125, 126
light, 7, 16, 66, 96, 107, 180

Index

light scattering, 7, 16, 66
lipid peroxidation, 104, 108, 109, 110, 111
lipids, 3, 30, 101, 110
liquid chromatography, vii, viii, 2, 4, 5, 6, 7, 13, 17, 52, 62, 63, 64, 66, 68, 70, 71, 73, 74, 75, 76, 77, 78, 80, 81, 82, 83, 84, 85, 86, 87, 88, 95, 167
liquid chromatography (LC), vii, viii, 2, 4, 5, 6, 7, 8, 10, 13, 14, 17, 18, 19, 20, 21, 22, 23, 24, 25, 26, 28, 29, 30, 33, 35, 37, 38, 39, 40, 41, 42, 43, 44, 46, 48, 49, 50, 51, 52, 53, 55, 57, 58, 59, 60, 62, 63, 64, 66, 67, 68, 69, 70, 71, 73, 74, 75, 76, 77, 78, 79, 80, 81, 82, 83, 84, 85, 86, 87, 88, 89, 95, 167, 183
liquid chromatography coupled to ultraviolet detection (LC-UV), 13, 55, 58, 65
liquid chromatography coupled with mass spectrometry (LC-MS), viii, 2, 5, 6, 17, 19, 29, 30, 33, 59, 64, 70, 74, 77, 78, 79
liquid chromatography-high-resolution mass spectrometry (LC-HRMS), viii, 2, 5, 6, 19, 35, 37, 38, 39, 42, 48, 49, 50, 52, 57, 58, 59, 67, 69, 70
liquids, 156, 157, 161, 165, 168
liver, 29, 74, 111, 134
low resolution mass spectrometry (LRMS), 5, 18, 19, 20, 26, 27, 29, 30, 31, 32, 67
Luo, 16, 71, 72, 80, 87, 167

M

management, 131, 134, 141, 143, 145, 147, 173
mass, vii, viii, 2, 5, 6, 7, 13, 17, 18, 19, 27, 28, 29, 30, 36, 41, 47, 50, 51, 52, 53, 55, 56, 66, 67, 68, 69, 70, 74, 75, 76, 77, 78, 80, 81, 82, 83, 84, 85, 86, 87, 88, 167
mass spectrometry, vii, viii, 2, 5, 6, 7, 13, 17, 18, 36, 48, 50, 55, 56, 66, 67, 68, 69, 70, 74, 75, 76, 77, 78, 80, 81, 82, 83, 84, 85, 86, 87, 88, 89, 167
matrix, 6, 11, 16, 36, 48, 51, 62
measurements, 36, 50, 51, 67
medical, x, 25, 131, 169, 170
medication, 130, 134, 176
medicinal plant, ix, x, 7, 10, 13, 28, 71, 72, 74, 80, 81, 86, 130, 131, 134, 139, 141, 142, 143, 145, 146, 147, 148, 149, 150, 151, 158, 160, 162, 165, 170, 176, 177, 180, 182, 184
medicine, 13, 130, 131, 145, 146, 147, 148, 151
medulla, 93, 94, 96
metabolic pathways, xi, 96, 170
metabolism, 101, 163, 164
metabolites, vii, ix, x, 3, 4, 5, 6, 9, 25, 35, 50, 73, 80, 92, 93, 95, 96, 97, 98, 99, 104, 105, 106, 107, 112, 115, 130, 139, 140, 142, 147, 148, 150, 151, 159, 160, 167, 170, 177
metabolomics, 5, 36
methanol, 7, 8, 9, 10, 12, 19, 20, 23, 24, 25, 26, 28, 40, 41, 42, 46, 48, 50, 66, 110, 136, 146, 154, 159, 160, 165
methodology, 63, 65, 167
mice, 111, 112, 164
microbial infection, 150, 151, 156
microorganisms, 3, 171, 173, 176, 185
mitochondria, 102, 103, 109
models, 17, 113, 163
molecules, 4, 5, 19, 101, 104, 119, 172, 179, 184, 185
morbidity, x, 132, 170
mortality, x, 132, 170
multidimensional, 65, 67, 88, 89, 179
multidrug resistance, 172, 175, 183
mutation(s), x, 111, 135, 147, 170

N

natural compound, ix, 11, 92, 105, 179
natural product, vii, 2, 69, 176, 180, 185
natural products, vii, viii, 2, 3, 4, 5, 6, 7, 8, 11, 14, 16, 17, 18, 20, 30, 36, 37, 38, 42, 49, 50, 51, 52, 54, 55, 56, 57, 59, 61, 62, 63, 65, 66, 67, 68, 69, 132, 140, 151, 162, 176, 177, 179, 180, 181, 184
neurodegenerative diseases, 101, 104, 108
nitrogen, 55, 112, 159, 163
NMR, 6, 7, 8, 10, 13, 15, 36, 38, 39, 52, 57, 58, 59, 60, 64, 66, 68, 72, 79, 81, 83
non-polar, 7, 30, 157
North America, 94, 114, 116, 117, 118
nuclear magnetic resonance, 6, 7, 82
nutrient(s), 92, 95, 143, 146, 150, 173

O

ODS, 9, 10, 18, 19, 23, 24, 26
oil, 150, 153, 167
optimization, 16, 63, 166, 167
organ, 133, 153, 179
organic solvents, 152, 153, 154, 158, 161, 164
organism, vii, 1, 2, 95, 154
orthogonal projections to latent structures-discriminant analysis (OPLS-DA), 10, 17, 25, 32, 39, 65, 67
overproduction, 101, 109, 111
oxidation, 104, 108, 110, 119
oxidative stress, vii, viii, ix, 91, 92, 101, 104, 108, 109, 110, 111, 112, 118, 122, 123, 124, 125, 126, 127
oxygen, 101, 102, 104, 108, 112, 121, 131, 139, 157
ozone, 112, 153, 158, 164

P

pain, ix, 129, 131, 133, 134
Panama, 40, 60, 83
Parmeliaceae family, 93, 122
partial least squares regression-discriminant analysis (PLS-DA), 8, 10, 16, 41, 47, 59, 61, 62, 67
pathogen(s), x, 92, 116, 159, 170, 173, 180, 181, 182, 183, 184, 185
pathway(s), 3, 5, 30, 96, 131, 171, 180
PCA, 8, 9, 10, 16, 21, 22, 25, 26, 32, 33, 34, 35, 39, 40, 41, 42, 45, 46, 47, 59, 60, 65, 67
penicillin, 134, 174, 179
peptide(s), 3, 4, 41, 47, 52, 53, 54, 62, 108, 123, 171, 177, 184
permission, 12, 15, 31, 33, 34, 49, 53, 54, 58, 60, 61, 64
peroxide, 101, 107, 108
pH, 7, 8, 24, 39, 41, 87, 106
pharmaceutical, x, 4, 27, 71, 134, 147, 160, 169, 176
pharmacological interest, v, 91, 101
phenol, 136, 148, 150, 151
phenolic compounds, 3, 7, 8, 9, 10, 11, 13, 20, 24, 28, 46, 50, 52, 63, 65, 78, 87, 105, 142, 159, 166, 167
phenols, vii, x, 107, 136, 137, 138, 150, 158, 159, 177
physicochemical properties, 4, 5, 6, 27, 35, 51, 66
phytochemical(s), v, vi, vii, viii, ix, x, xi, 1, 2, 3, 5, 23, 26, 28, 32, 33, 35, 60, 69, 70, 72, 73, 78, 79, 80, 83, 119, 129, 130, 134, 136, 137, 138, 139, 141, 142, 143, 144, 145, 146, 147, 148, 149, 150, 151, 158, 159, 161, 162, 165, 167, 169, 170, 172, 173, 176,178, 180, 182, 183
plants, vii, viii, ix, x, 2, 7, 10, 19, 25, 27, 30, 51, 71, 73, 74, 75, 79, 95, 130, 132, 134,

135, 136, 137, 138, 139, 140, 141, 142, 143, 144, 145, 146, 148, 149, 150, 151, 156, 158, 159, 160, 162, 165, 170, 173, 176, 177, 184
playing, 36, 37, 51, 59
PLS, 8, 9, 10, 16, 38, 41, 43, 47, 59, 61, 62, 67
polar, 7, 11, 18, 37, 44, 47, 66, 145, 152, 157
polarity, 47, 151, 153
polymerization, 131, 139, 140
polyphenol, 79
polyphenols, x, 9, 10, 14, 25, 26, 70, 72, 166, 170, 177
population, ix, 129, 130, 131, 132, 176
preparation, 27, 29, 77, 151, 161, 171, 181
prevention, vii, 2, 3, 66, 104, 112, 142
principal component analysis (PCA), 8, 9, 10, 16, 21, 22, 25, 26, 32, 33, 34, 35, 39, 40, 41, 42, 45, 46, 47, 59, 60, 65, 67
principles, 135, 155, 180
proteins, 95, 101, 108, 110, 140, 164, 174
public health, 171, 172, 180
purification, 56, 151, 152, 153, 157, 180

Q

quality control, 13, 27, 28, 84
quantification, 7, 11, 12, 14, 16, 27, 28, 30, 74, 75, 76, 77, 84, 85, 86, 87, 146
quercetin, 32, 33, 104

R

radicals, 104, 106, 110
ramp, 40, 46, 57
RBC, 131, 135, 139
reactive oxygen, viii, 91, 101, 107, 112
recognition, 83, 134, 172
researchers, 109, 134, 159

resistance, vii, x, xi, 139, 151, 161, 169, 170, 171, 172, 173, 174, 175, 180, 181, 182, 183, 184
resolution, viii, 2, 5, 6, 11, 18, 30, 36, 42, 43, 44, 45, 46, 47, 50, 51, 52, 54, 55, 56, 57, 63, 66, 67, 69, 80, 81, 82, 83, 84, 85, 86, 87, 89
response, 29, 51, 78, 144, 167
reversed-phase liquid chromatography (RPLC), 7
risk(s), 134, 150, 154, 164, 172
root, 3, 28, 72, 142, 143, 146, 147, 163, 177

S

safety, 27, 130, 151, 152, 154, 155, 157, 179
saponin, 135, 136, 137, 138
scope, viii, 2, 6, 55, 69
secondary metabolites, vii, ix, 3, 4, 5, 6, 35, 80, 92, 95, 96, 97, 98, 99, 104, 105, 106, 107, 112, 115, 130, 139, 140, 142, 147, 148, 150, 151, 159, 160, 177
seed, 53, 64, 65, 140, 142, 146, 147
selectivity, 7, 16, 18, 37, 67
sensitivity, 5, 6, 14, 18, 28, 32, 37, 51, 67, 87
shortness of breath, ix, 129, 131
showing, 14, 15, 17, 50, 58, 59, 61, 112
sibling, 133
sickle cell disease, vii, ix, 129, 130, 141, 142, 143, 144, 145, 146, 147, 148
side effects, 131, 134, 135
signal transduction, xi, 111, 170
similarity analysis (SA), 16, 17
software, 16, 17, 53, 179
solubility, 4, 153, 157, 163
solution, 7, 8, 21, 27, 46, 87, 105, 152, 154, 177

solvents, ix, 19, 41, 130, 149, 152, 153, 154, 155, 156, 157, 158, 159, 160, 161, 162, 163, 164, 165, 166, 167, 168
South Africa, 58, 79, 80
Spain, 1, 68, 91, 94, 117
species, viii, ix, 3, 4, 28, 32, 35, 40, 60, 61, 64, 72, 78, 79, 80, 83, 91, 93, 94, 95, 96, 97, 101, 104, 105, 106, 107, 109, 112, 129, 139, 153, 171, 176, 178
spectroscopy, 6, 13, 52, 60, 64, 82
stability, 11, 27, 55, 82, 158
steroids, 26, 43, 76, 86, 96, 135, 136, 137, 138, 139
stress, vii, viii, ix, 19, 75, 91, 92, 101, 104, 108, 109, 110, 111, 112, 123, 124, 125
structure, 14, 32, 60, 96, 120, 159, 160, 174, 175, 177
substrate(s), 94, 108, 153
Sun, 72, 77, 84, 85, 86, 126, 167
suppression, 51, 87, 88, 111
symbiosis, viii, 91, 92
syndrome, 52, 88, 133
synthesis, 92, 96, 154, 157, 165, 171, 179

T

tannin(s), ix, x, 3, 50, 65, 130, 135, 136, 137, 138, 148, 150, 151, 170, 177, 178
target, 18, 27, 31, 48, 51, 80, 112, 166, 172, 173, 174, 175
techniques, viii, 2, 5, 6, 14, 27, 36, 51, 62, 63, 66, 67, 95, 105, 134, 163, 166, 171, 179
technology/technologies, x, 140, 149, 161, 179, 180
temperature, 16, 106, 155, 157, 158

therapeutics, x, 143, 169
therapy, 130, 154, 176, 179, 184
threats, 151, 162, 172
toxicity, x, 149, 153, 154, 155, 156, 158, 162, 163, 164, 165
transplantation, 133, 140, 143
treatment, ix, x, 16, 19, 28, 36, 111, 112, 129, 130, 133, 139, 141, 142, 145, 147, 150, 169, 173, 180
trifluoroacetic acid, 9, 10, 11, 12, 19, 22, 66
two-dimensional liquid chromatography (2DLC), 5, 62, 63, 64, 65, 68, 70, 88

U

United States (USA), 68, 69, 82, 132, 165

V

variations, 7, 18, 132
vegetables, 4, 121, 160, 177
versatility, 50, 66, 67
viruses, xi, 170, 177

W

water, 7, 8, 9, 10, 11, 13, 19, 20, 21, 22, 23, 24, 25, 26, 38, 39, 40, 41, 42, 43, 44, 45, 46, 47, 48, 49, 50, 52, 53, 66, 92, 93, 102, 104, 131, 136, 137, 152, 153, 159, 160, 161, 168, 171
World Health Organization (WHO_, 150, 151, 153, 164, 172, 176, 178, 182, 183, 184, 185
worldwide, viii, 4, 91, 93, 156, 173